C000143942

1,000,000 Books

are available to read at

Forgotten Books

---◆---

www.ForgottenBooks.com

---◆---

Read online
Download PDF
Purchase in print

ISBN 978-1-332-14361-0
PIBN 10290550

This book is a reproduction of an important historical work. Forgotten Books uses
state-of-the-art technology to digitally reconstruct the work, preserving the original format
whilst repairing imperfections present in the aged copy. In rare cases, an imperfection in
the original, such as a blemish or missing page, may be replicated in our edition. We do,
however, repair the vast majority of imperfections successfully; any imperfections that
remain are intentionally left to preserve the state of such historical works.

Forgotten Books is a registered trademark of FB &c Ltd.
Copyright © 2018 FB &c Ltd.
FB &c Ltd, Dalton House, 60 Windsor Avenue, London, SW19 2RR.
Company number 08720141. Registered in England and Wales.

For support please visit www.forgottenbooks.com

1 MONTH OF
FREE
READING

at

www.ForgottenBooks.com

By purchasing this book you are eligible for one month membership to ForgottenBooks.com, giving you unlimited access to our entire collection of over 1,000,000 titles via our web site and mobile apps.

To claim your free month visit: www.forgottenbooks.com/free290550

* Offer is valid for 45 days from date of purchase. Terms and conditions apply.

English
Français
Deutsche
Italiano
Español
Português

www.forgottenbooks.com

Mythology Photography **Fiction**
Fishing Christianity **Art** Cooking
Essays Buddhism Freemasonry
Medicine **Biology** Music **Ancient**
Egypt Evolution Carpentry Physics
Dance Geology **Mathematics** Fitness
Shakespeare **Folklore** Yoga Marketing
Confidence Immortality Biographies
Poetry **Psychology** Witchcraft
Electronics Chemistry History **Law**
Accounting **Philosophy** Anthropology
Alchemy Drama Quantum Mechanics
Atheism Sexual Health **Ancient History**
Entrepreneurship Languages Sport
Paleontology Needlework Islam
Metaphysics Investment Archaeology
Parenting Statistics Criminology
Motivational

INDUSTRIAL

EXPLORINGS

IN AND AROUND LONDON.

BY

R. ANDOM,

*Author of "We Three and Troddles," "The Strange Adventure of Roger
Wilkins," &c., &c.*

WITH NEARLY ONE HUNDRED ILLUSTRATIONS BY
T. M. R. WHITWELL.

London :
JAMES CLARKE & CO., 13 & 14, FLEET STREET.
1895.

G 8 B2

AFTER GIVING
THE MATTER DUE
CONSIDERATION, AND RECOG-
NISING THE FACT THAT NONE
OTHER CAN SO ADEQUATELY ESTIMATE
AND APPRECIATE THE PAINS AND PENALTIES AS
WELL AS THE PLEASURES AND THE
PROFITS ATTENDING THESE EXPLORA-
TIONS, I HAVE CONCLUDED
TO DEDICATE THIS
BOOK TO MYSELF.

1091463

PREFACE.

In days of yore two men rode into a certain market-place from opposite directions, and it chanced that they halted on either side of a shield that was set up on high there, though

what earthly use it could serve in that position I am at a loss to understand.

"Good morning," said the first traveller,

pleasantly. "Charming weather for this time of the year."

"Lovely," rejoined the other. "I was just admiring this quaint old silver shield. Can you tell me what it signifies?"

"Well, I should say it chiefly signifies that you are an ass to take a bit of wrought iron for sterling silver," retorted the other, brusquely. I expect he was a humorist.

And then they proceeded to argue the matter

in the blunt, old-fashioned manner that inspired respect, even if it did not always carry conviction.

As a matter of fact, the shield was of brass, tarnished on one side and tinned on the other.

Now I also have set up a shield in the market-place, and I want you all to ride out in your thousands and buy it, and read, mark, learn, and inwardly digest it, and make presents of it to your friends, relatives, and acquaintances, and tell everybody what a really excellent shield it is. Only beware! It is a two-sided shield; and lest you should go astray over it, I will tell you that it has the silver of humour on the one side, and the iron of hard fact on the other, and if you drill through you may come upon some solid metal of good sterling quality underneath. If you do, I hope you will let me know. It will surprise me rather; but the gratification I shall obtain will more than compensate me for the shock.

R. ANDOM.

WOODFORD,
        August, 1895.

CONTENTS.

CONTENTS

PROLOGUE.

THE travelling instinct has always been more or
less strong upon me, and at one time I used to
look forward to the day when I should spurn

the sand of Sahara's desert, which makes good,
soft spurning material I have been led to

understand, and tread mighty continents and untoured lands. As a matter of fact, I suppose, when the time came, I should do the continental treading in a railway train, as being less tiring and more practicable; but that is the way I used to think about it and dream about it. We never realise our ideals on this earth. The nearest point I have as yet reached to the Sahara is Jersey, and anything between Edinburgh and the North Pole is untraversed ground so far as I personally am concerned.

But quite recently the compensating balance of life, which maintains the world's equilibrium, indicated a sphere wherein I might exercise my bent, on a somewhat contracted scale, truly—a sort of parlour-billiards, so to speak—but with pleasure to myself and profit, I trust, in all that humility which is so conspicuous a feature in my character, to my readers. That same compensating balance, by the way, is a wonderful dispensation. Without it one half of the inhabitants of the civilised earth would be lunatics, made so by over-joy, and the other half would be suicides driven thereto by grief, vexation, and disappointment. With it in active operation there are only isolated cases of such melancholy terminations, which have been overlooked and forgotten, perhaps, in the press of business, for the compensating balance has a very busy time of it indeed. But ordinarily, when a man is getting too prosperous, too wealthy, too great, and begins to show signs of a swelled head, which is the first stage to joy-lunacy, up goes the lever, bang goes a bank or something, off goes his yellow jacket, and there he is, sane and sensible, albeit a trifle savage, once again. And so in the reverse cases grief is turned to joy, and the suicide is postponed.

But I started to tell you about my exploration scheme, limited—very limited indeed, I may say, seeing that it was designed to be confined within the twelve-mile radius.

It was in this way. I was reflecting—I do a lot of reflecting at odd times; it prevents that sense of idleness when you have nothing to do. I don't mind being idle when I am simply hemmed in with work; in fact, I rather enjoy it. But when I have nothing in hand, I hate to remain a drone—it is a peculiarity in my composition—and so I reflect.

"I cannot be a traveller," so ran my reflections, "because I can't get away from London. No one can expect to be a satisfactory traveller, and go about the world and see things, who has to put in an appearance in London not less than three times a week. It hampers him, so to speak, and prevents him from giving his full attention to all the lovely and curious things he has to look at and describe.

"But why go away?" I went on reflecting. "Where is the necessity? If you can't go to

Kamtschatka, go to Kentish Town and describe that. Failing Bermuda, why not travel to Battersea? There are a lot of lovely and curious things in Battersea! And if the Solomon Islands are 'off,' why not let Stratford be on? There are some interesting savages about Stratford

who could give points to any child of nature roaming in insufficient clothing across the coral strand of the Solomon Islands. Be a patriot! Cry up your own country! Write about your own people—your own native productions! It is easier than going abroad to do somebody else's, it is safer and more comfortable, and it is cheap." I saw my opportunity—nay, my duty—and, as Lindley Murray would say, I done it.

INDUSTRIAL EXPLORINGS.

IN PIANO-LAND.

T is, generally speaking, best to start an enterprise with something big. I started with pianos, and took an overcoat, a pipe, a newspaper, and a threepenny ticket, and sallied forth on my travels. And, I venture to think, never was expedition more easily, expeditiously, and economically fitted out and despatched.

I went to Piano-Land, and I went there direct. If I had been Henry M. S——y I should have gone to a jam factory at Blackheath and struck across country and discovered Piano-Land for myself; but as the Midland

Railway had already located the place, I resolved to leave them the *kudos*. Privately I don't
think it is very much to their credit, though, of
course, when they first found it it may have
been pretty and picturesque enough. I found
it depressing and more than a trifle dirty, and
worn down at the heel, so to speak.

It was dull and cold, and foggy, too, when I
left Kentish Town Station in search of Grafton
Road and promptly and immediately lost myself.
In answer to my pathetic inquiries several
kindly-disposed persons did stop and give me
minute and precise information whereby I
might travel to my destination; but they im-
part topographical information in a style of
their own up Kentish Town way, and I hadn't
the key. Instead of telling you, for instance,
to take the first turning to the right, and the
fourth to the left, and then keep straight on,
they say: "Grafton Road? Brinsmead's place?
Why, that's where they makes pianos!"

You murmur your interest in the information,
and signify that you have no vital objection
thereto provided the work is carried on on
sanitary principles.

"Want to go there?"

You casually observe that you *would* like to stroll in that direction if it is quite convenient. Can you be directed?

"Yus! Go straight along till you comes to Jones's the butcher's, then turn round by a row of red-bricked 'ouses, and you'll see the 'Cow and Mustard Pot' on the other side of the road, and just past that there's a turning, and there you are."

And you go on merrily, and find Jones's the butcher's, and the "Cow and Mustard Pot," and the red-bricked 'ouses, and fancy your travels are nearly ended. Getting a trifle uncertain, perhaps, you ask again.

"Grafton Road, sir? There ain't no Grafton Road about here! There's a Clifton Road."

My enterprise was saved from utter extinction at the outset by a dirty little boy who was playing with his "sucker"—a leather and rather a messy plaything, which has a moral influence on the young, I am told, besides illustrating several well-known natural laws plainly for their intelligent comprehension.

I asked him what percentage of the wealth of the Indies would induce him to personally pilot me to Grafton Road. The boy was prompt at a

deal; he said he thought "tuppence" would do
it, and I closed on the bargain, and told him to
lead on.

Ten minutes later, I stood before the imposing
blocks in which Messrs. Brinsmead and Sons
carry on their manufacture. A gate within a
gate admitted me, and I was in Piano-land—
and a very wonderful and interesting land I
found it. A land overflowing with glue and
timber and tinkling harmony, colonised by its
own people, shut in from the outside world
by a high wall, and jealously guarded by
massive gates, through which only the specially
privileged pass. I presented my credentials to
Mr. Thomas Brinsmead, who came out to me,
and he, courteously brief and pleasantly
business like, handed me over to the charge of
Mr. Hall, his foreman, under whose very able
guidance I made my journey. It is the fashion
to be personally conducted nowadays. Some
really great explorers adopt it, so I felt no
shame, albeit it was rather prosaic.

Mr. Hall would not make a bad journalist,
so quick is he to grasp the explorer's wishes.

"Don't let us be technical," I said. "Let
us walk round and spy out the land, so that I

may perchance write a big book on the subject; and if we meet any spades about anywhere let us call them shovels — just simply plain

PREPARING THE FRAMES.

shovels." And we were pleasantly chatty and descriptive thereafter—at least, Mr. Hall was. I had my reputation as an explorer to maintain, and I dared not unbend too far and become jocose, lest peradventure he discovered that I was a humorist, and not a real, *boná-fide* traveller at all.

There is only one feature that an outside view of the Grafton works permits you to retain, and that is the monumental piles of timber that are reared up everywhere, even on the roof-tops.   Stacked with mathematical precision, each plank kept apart from its neighbour by short cross-ties, to allow the air to circulate freely in between ; black with age and weather—and I should judge the weather to be pretty thick occasionally around Kentish Town—these planks, gathered in from all parts of the world, wait their turn to be converted into " Uprights " and " Grands."   A very lengthy wait it is, too.   Four years, on an average, between the time of stacking and actual use—the last nine or twelve months under cover, in a universal temperature of eighty degrees winter and summer—are allowed to elapse to thoroughly and completely dry all trace of moisture out of the wood.   Moisture causes trouble in pianos, and every care is taken to exclude it.   Therein the instrument differs from some performers I have known, who have displayed a remarkable aptitude for moisture without thereby affecting their tone or touch, and have not limited the quantity either, though I have heard them grumble at the quality.

Thoroughness is strikingly in evidence all through the various departments, and presumably Messrs. Brinsmead hold that it is better to overdo a matter than to take chances. And undoubtedly they are right. I have known men who have gleefully speculated in instruments made in the Faderland, and have come to wish they hadn't. They have found——but there, you know the story, I daresay. Anyway, it is not pleasant to have a thirty-guinea piece of property that you designed for use as a piano develop such a guileless, open-hearted, frank disposition, that it feels impelled to open its casing and display its interior economy to the passers-by, just to show that there is no deception, and that it really does contain strings and fittings. Nor does the ambitious variety altogether please. I mean the sort that rebels at the comparatively simple and unobtrusive *rôle* of piano, and splinters itself up into wash-tubs and kitchen stools, with perhaps a little surplus material left over for a window-sash and a chicken-coop.

Here, perhaps, I ought to interject a word or two of praise about Messrs. Brinsmead's instruments; but I will not. They are of world-wide fame and speak for themselves, if you know how to play and can make them, and they haven't any ideas above their station like those I have alluded to.

From the wood-stacks—which is the beginning of all things, as far as pianos are concerned—I was conducted through a series of rooms where the treatment of the iron frames was in progress. Round about the walls, and in odd corners, these frames, rough and unfinished, as they had come from the foundry, were piled, while the various operations which reduced them to a finished and ornamental condition were in full swing. Here might be seen a number of men rubbing down, with infinite toil and patience, the rough ironwork by a sort of "holy-stoning" process; while away off in a far corner a "hand" was viewing, with a look of conscious pride, a magnificent great plate destined for a "Grand,"

that shone like burnished gold. I walked across
to it and inspected it closer.

"Enamelled and baked," Mr. Hall remarked,

DRILLING HOLES FOR PINS.

briefly. And I understood that I might safely
handle it without the colour coming off in the
process.

In another building vertical drills were at
work on these same plates, boring out the holes
for the pins. The plates slid along under the
deft guidance of a mechanic—rattle, whirr,
scrunch, and the hole was bitten out as clean
and easily as the cheese-taster's knife pierces a
new importation of Cheddar.

In a big room adjoining I came upon a real
land of marvels. There was a clatter and hum
of powerful machinery, a pleasant smell of
fresh wood—and it is astonishing how fresh
and scented some of these woods do strike the
nostrils—and sawdust lying about by the sackful, and floating round in clouds. I could have
passed for a very respectable miller before I
had been in the room ten minutes—that is, if
any one could conceive a sort of fancy-dress
miller with a tall hat and a note-book.

Talk about the little busy bee! Any specimen hive would have gone out of business in
sheer envy and impotence could they have
peeped in and seen that scene. Sawing—in
which operation, by the bye, both the fussy little
"circulars" and their bigger and, if possible,
more fussy prototypes, and the dignified and
somewhat sinister-looking "band" saws are

concerned — planing, moulding, and shaping were going on everywhere, and no pen can adequately give one's impressions on seeing a huge, roughened plank slid on to one of the massive steel beds of the planing-machines, and come through, smooth and glossy, and pared down to a standard thickness.

The point that most struck me, after I had travelled a little farther, was the slap-dash, easy-going, almost contemptuous way in which the various parts are turned out, in contrast with the precision and nicety displayed when it comes to fitting those parts together. To make the sections of a piano is easy—if you know how; I felt that I could do it myself—if I knew how; but I should despair of ever bringing the parts together to a harmonious— literally speaking—whole. The Chinese puzzle is not a circumstance to it.

A very open sort of lift, which connects the various floors, took us to the top of the building, and the peaceful quiet, after the whirr and rattle below, was striking.

On these upper floors, cases for " Grands," sounding-boards, and backs, with their exceptional construction of " Wrest Plank," by which no two strings pull on the same grain of the wood, are variously constructed, and important and interesting operations they all are.

The sounding-board, for instance, is made in sections, each section being matched to its fellow, and the whole glued together diagonally. The nicety with which it has to be proportioned in thickness, combined with the care necessary in adjusting the "bridge" over which the strings pass, renders it perhaps the most important of details that are all important.

Hard and beautifully white before they have received the preserving coats of varnish, these sounding-boards are marvels of constructive skill and ingenuity. We were standing against one that had just been fitted into a "Grand," while the process of construction was being explained to me, and Mr. Hall struck it—the board I mean, not the process—with his hand. It sounded as deep and resonant as a big drum. I paid my tribute of admiration and walked forward, for I was afraid my stock of admiration tribute would give out under such frequent calls and leave me stranded. Vainly! The next call came a moment after, and a heavy draft it was.

Glue, which is a prominent feature in these works—and a very harmless, inoffensive, and

THE BAND SAW.

unobjectionable feature it is, too; quite contrary to any similar feature that I had hitherto

encountered, and without that " smelly " peculiarity so noticeable in glues—plays a very
important part in fitting pianos together. And for
a successful gluing job there must be pressure
and warmth. The warmth here, as elsewhere, is
obtained from steam-heated cupboards. The
pressure is given by a simple but ingenious
method that quite took my fancy.

"Whatever are those things for ? " I queried.

My guide laughed. "I thought that would
interest you," he remarked. "It usually does.
They are ' go-bars.' "

I overlooked this lapse into technicality for
the sake of further information, and was re-
warded for my forbearance.

These " go-bars " are lance-wood staves such
as they used to fashion into bows for bold
Robin Hood and other irresponsible persons
who objected to taxes and convention. The
parts to be glued together are fitted in position,
and one end of the " go-bar " is placed upon
them, the other finding a rest against a false
roof a few feet above. A pressure amounting
to tons, resulting from the attempt of the staves
to straighten themselves again, can thus be
brought to bear on the glued surface.

In making the cases for "Grands"—an operation I before alluded to—glue and pressure play a considerable part. The cases, whether of "concert" or "boudoir," are composed in layers glued together, these having been previously steamed and bent round into shape by huge clamps; and as every one is familiar with the completed article, a due pondering over this fact, with a little mental exercise, will enable its significance to be borne in upon one.

Hard as iron and rigid as steel they come out of the clamp-frame ready for the fitter's deft handiwork, and by no cutting, piecing, or contriving can a similar result be arrived at.

On every floor there are store-rooms containing the fittings that will be required, for a piano is not made as one generally understands the term; it is built up by many processes, and in many departments. The parts, as I before explained, are made wholesale, and are delivered in sets to the "fitting" shops.

Tinkling sounds were by this time to be heard from various quarters; not unmusical nor grating, as one might expect, but suggestive of harps being touched by unskilled hands. It pleased me, for it seemed in keeping with the place. Disjointed and jerky it perhaps was, and lacking in the grace and finish of a Beethoven sonata; but it was a suggestion of melody anyway.

Art and music run closely together, and they did so then, for with the ear-pleasing quality come the gratification of the sight; and a Brinsmead piano can charm both senses. Panels delicately inlaid—I could have sworn that they were hand-painted, and beautifully done at that—and veneers, polished and natural, constituted the elements that gratified my artistic sensibilities.

I was shown, somewhere or other about the building, these veneers, all cut and pieced together ready for use.

Sliced, from warty burrs on trees growing in Persian forests, into thin strips that are as fine and as pliable as paper, albeit they are tough and workable, the patterns are matched with admirable dexterity. The sheets are then pasted together temporarily into panels. A glue bath gives them the consistency and pliability of a fine leather, and pressure suffices to face a panel or a moulding so neatly and

FIXING THE "GO-BARS."

accurately in pattern and join that detection is quite impossible.

"It is not cheapness; it is necessity,"

2

remarked Mr. Hall, as we viewed the process, and I, understanding him, forebore comment.

The method of veneering a moulding is particularly ingenious. A lengthy metal matrix receives the moulding with the facing downwards, the inevitable pressure is applied, and presto! you are gazing at a length of beautifully patterned walnut.

Somewhere up under the roof the frames are strung, and they are tuned and tuned and tuned, and still they go on tuning. They seem to have a liking for the job up at Grafton Road.

"Hello!" they say, "here's a piano. Let us tune it." And they do so as long as the instrument remains in sight. I believe the carriers have to take extra hands with them when they go to fetch a piano away, to head off any over-zealous tuner, while the remainder are getting it into the van.

You can't get away from notabilities. We were walking past a beautiful specimen of a concert-grand, that was lying on the stocks in a very incomplete state. It was being built, I was informed, to the order of Mr. George Grossmith. Several fine old instruments, though battered by storms and soaked by the salt waves, were pointed out to me, too, in my peregrinations. They had travelled to all parts of the world, and had come up there for rest and refresh——repairs, I mean, from the saloons of the P. & O. Co.'s boats. They looked as though they wanted it.

"That is a good bit of felt," Mr. Hall observed, as we passed into a small off-room, lined on one side with cupboards. He handed me a solid-looking square of material white as snow—driven snow is the correct phrase, I believe—which might have been anything from a patent dog-biscuit to a bath-mat.

I felt it and agreed with him. Felt plays a considerable part in the interior economy of pianos. It is used in protecting, and "checking," and it covers the hammers, so that when they strike the wires they may make themselves—but there, I will spare you such an obvious piece of gratuitous flippancy.

The woodwork of the hammers is first cut through, and then a length, containing I don't know how many because I didn't ask, is wrapped round with the felt casing, which is glued on the inside. Now, if you look at one of these hammers you will observe that it is

thickest at the top and fines away as it gets
round towards the foot. This is done by hand,
wholesale, by a kind of sand-papering process,
which rubs away the superfluous material
to the required standard, and then each covering is clinched on to the wood with a metal
pin, the length is divided up into sections,
and the hammer is ready, or rather a whole
batch is ready, to be added to the general
store.

And so it came to pass that I found myself
on the ground-floor again, having travelled the
length and breadth and the height of the firm's
premises. I had passed by hundreds of pianos
—in fact, I had come to look upon them as
commonplace sort of articles, rather in the
nature of litter—and I had witnessed the
inmost details of their construction; and they
are many and marvellous are these same
details, for there are more parts in your piano
—supposing that you have a piano—than you,
with all your philosophy, dreamt of, John—or
Evelina, if it be a lady whom I address—
unless you have an enlarged mind, and would
set the number at six thousand odd in the first
instance.

It is impossible to do justice to all these matters, and there are many points that I have forgotten, or been compelled, reluctantly enough, to forego. For instance, there is that of pinning for the strings, the tap, tap, tap, of ceaseless hammers driving in hundreds of steel pegs to guide the strings, with noise enough to wake all except the real, *boná-fide* defunct.

Then there are the operations of regulating and adjusting the action, pedalling, and polishing, to say nothing of the ingenious and economical devices by which such satisfactory results are obtained. One of these I must mention, it is so instructive an instance of economy of waste product. The exhaust steam from the engines, which is ordinarily blown off into the air to add to the miseries of countless thousands, is here carried through a system of pipes running on every floor, to boil the glue, warm plates and materials, and to make itself generally useful.

For what I have set down I will hold myself responsible; but that no aspiring amateur who had designed to make himself a piano in his spare time from my paper—wood would be a better material—may deem himself defrauded,

I will give a brief *résumé* of the process which I hope will meet the case. Here is the recipe. First let the amateur get his material, and cut it, and plane it, and make it look pretty. Then he must make the back, and the sounding-board, and the casing, and stick them all together—tenpenny nails will answer and are cheapest; but glue and screws look better. After that put in the mechanism, add whatsoever seems necessary, polish and decorate, and present to the nearest deserving institution for the deaf and blind.

After a brush down, which I sorely needed, I parted from Mr. Hall, in a snug little office just within the gates, with a due acknowledgment of his kindness, that had rendered my first exploration so easy, interesting, and pleasant; and quitting Piano-Land with a sigh of regret, I made my way, unguided this time, to the railway-station, and got home in time for tea.

IN ROPE-LAND.

THICK FOG, as dark as Erebus and as pliant as the tide of the silvery Thames off Blackfriars, wrapped the East-end of London in a grimy shroud when I stepped out of the train at Coborn Road in quest of Rope-Land.

Fortunately, Rope-Land lay just behind the station, and I was spared a repetition of topographical misadventures. This was fortunate indeed, for they would have been multiplied and intensified beyond my limits of endurance under such conditions as then confronted me.

Every traveller out of Liverpool Street over the Great Eastern system is acquainted with the outward aspect of Rope-Land. He will note the various buildings and sheds, the long

" walk " which runs—a singular contradiction
in terms—parallel for nearly a mile with the
rails; the ingenious condensing arrangement
by which the water used for cooling the exhaust
steam is conveyed along the top of this same
building, and eventually, after playing in a
series of waterfalls over the roof, for the
purpose of lowering its temperature, finds its
way back to the boiler-house again, to be
reconverted into steam—a device economical
and praiseworthy from everybody's point of
view, except, perhaps, the water company's.
He may even hear the rattle of machinery, and
the ponderous thump-thump of the engine
below vibrating through the station above.

A comparatively narrow frontage has this
Rope-Land of ours, for it runs to length rather
than breadth; and a flat succession of windows,
a private door admitting to the offices, and the
inevitable big gates all bearing the legend,

J. T. DAVIS,

ROPE AND TWINE MAKER,

presented themselves, a bit at a time, owing to
the fog, as I made my way round through the

METHOD OF DRIVING.

yard to be received
by the gentleman—
I, or rather the rail-
way company, had
kept him shamefully
waiting, by the bye—
under whose guidance I made my tour of
inspection.

In olden days—and I believe Mr. Davis

himself was in business in that way at the
time, for Rope-Land is a land of slow growth
and long development—a man stood at one end
of the yard and a boy at the other, and they
made a rope between them. "They don't do
that sort of thing now," as the song says.
To-day it is a marvellously perfected work,
involving spacious premises, a big plant, and
costly machinery, so huge and powerful, yet
intricate withal, that I lost myself in delighted
astonishment at the outset, and remained lost
for the remainder of the afternoon.

First, I was taken into the engine-house, to
whet my curiosity perhaps; anyway, it had
that effect. Imagine for yourself a huge
structure, gleaming as only skilful and conscientious engineers know how to make their
steel and brass work gleam, the high and low
pressure cylinders cased in dark, shining
mahogany, and the ponderous driving-wheel,
half underground, measuring sixteen feet in
diameter, and weighing seventeen tons, all so
beautifully balanced that, save for the thump,
thump, thump previously alluded to, it is
almost inaudible. Some little one-and-a-half
horse-power gas-engines I have been acquainted

with could give it fits in the matter of noise.
And yet for all that it supplies the whole of
the motive-power in the works, although it is
pretty severely taxed in extra busy seasons to
"come up to time" with the credit that has
characterised its five years' service. Its pre-
decessor, looking a veritable pigmy, still finds
an honourable home behind it, and "earns its
oats" by operating the "rope-walk" machinery
when holidays and other irritating interrup-
tions interfere with the more extended operations. Both these engines are supplied by
two boilers on the same extensive scale, which
have their home in a building on the other
side of a narrow pathway laid down with rails
for the convenient transportation of heavy
materials.

To return to the engine-house. The driving-
wheel is ribbed on its face, very much like a
piece of corrugated iron, and transmits its
power by a series of manilla ropes, made, of
course, on the premises. These, I gathered,
stand easily the strain that could not be put
upon ordinary belting, are very satisfactory,
and very economical by comparison. The same
system maintains throughout the works. Ropes

and corrugated wheels take the power from building to building, geared to suit the special requirements, as shown in the illustration.

Electricity plays its small but very important part in this house. In every building where machinery is running an ordinary electric push is fixed, the wires running into the engine-house and connecting with a throttle device. If an accident or some untoward event takes place, no matter where, this button is pressed, and presto!—no joke intended!—the arrangement is set in action, and the huge monster comes to rest. After I had seen some of the machinery I began to appreciate this handy little arrangement.

Away up on one side of the wall, another contrivance of pipes and plates, and a pressure gauge, attracts the attention. It is the business end of an automatic fire-brigade, which is calculated to make any ordinary human fire-brigade feel sick of futile envy and chagrin. Hemp, and manilla, and the other products being dealt with are dangerous commodities where fire is concerned, and to discount these risks as far as possible, automatic sprinklers are fixed everywhere, so that in times of need

every square inch of space receive its due
allowance of water. These sprinklers are

THE CRUSHER.

sealed with a soft metal, which fuses at a
certain temperature, the air which has been

pumped through the system to hold back the water finds a speedy exit, and is followed up as speedily by a deluging stream, and the fire, concluding that it has mistaken the day or the address, apologises and goes out.

Noting these facts, I left the engine-house and made my way over the rails and through the fog, past tons of raw material in the shape of manilla, jute, hemp, and fibre, which is landed at Coborn Road in canvas and iron-bound bales, just as it comes home from Manilla, New Zealand, and the Indian ports— mostly Bombay and Calcutta.

They are curious-looking substances—various in themselves, and varying still more in qualities : some as white and soft as cotton and hair, others rough, coarse, and tough, and seemingly impracticable.

My guide and an old Scotch engineer between them contrived to make me understand the process by which this unpromising-looking, shock-headed tangle of material became anything, from coloured twine to a ship's cable, and I will endeavour to transmit as much of it as I can remember, and make myself clear about, without the aid of

A "CARD," OR SHREDDING MACHINE.

the paraphernalia they had to illustrate their meaning.

The very first step in the operation brought

me to a long horizontal contrivance that looked capable of giving points to any torture-chamber fitting or accessory ever devised; in fact, Rope-Land contains as many wickedly fiendish-looking machines as it has ever been my lot to witness. This particular one was a sort of mangle arrangement, having a number of curved and ribbed rollers, as much like the travelling knives of a grass cutter as anything except that they are big and solid and blunt. There is a double row of these, and they revolve within each other in reverse directions, so that when the hemp or jute, or any material hard and knotted enough to require it, is sent travelling down that terrible mangle, it is so pressed and crushed that it comes out at the other end in a softened and workable condition. During its passage an overhead contrivance damps it with a mixture of water and oil. This is only a preliminary discipline, however.

Round about the floor-space in two of the buildings several huge structures, resembling threshing machines somewhat, are grouped. These are technically known as cards, and their interior economy is a sight to see. A huge roller in the centre, with several smaller

A TWISTING MACHINE.

editions of itself round about it, revolving in reverse directions, and all bristling with short steel points as close as they can be studded in,

3

is what a removal of the casing showed me.
Ordinarily, they are kept religiously boxed up,
for once caught by those vicious-looking teeth
and nothing can save the substance or material
from being drawn in and shredded into the
finest fragments. That is its mission, and like
every well-regulated, right-minded piece of
mechanism, it is content to perform that
mission blindly. Whatever is given it, it
doubtless argues, is given it for shredding
purposes, and if they choose to throw a
machinist in occasionally—well, it is tough
work, but duty is duty! So the powers that
have control accept its blind, unreasoning
service and fence it in carefully, so that it may
not have its feelings harrowed by uncongenial
occupation.

Into this huge box of gleaming teeth, which
are set diagonally on their respective barrels,
the crushed and prepared material is fed, and
the gentle combing process that it then experi-
ences can be easily imagined. Each roller is
geared to make so many revolutions a minute,
and the slowest set accomplish something like
two hundred and fifty.

A curious result after the nature of fireworks

is obtained, I was told, when a bit of grit or wire or such-like foreign substance accidentally finds its way in with the jute or hemp; but it is an expensive and risky experiment, and one not to be undertaken for the mere gratification of an exploring scribe, and so I had to rest content with an outline description.

Some of the finishing cards are marvellous examples of engineering constructive skill and ingenuity. In following the rope-making process I came upon one, a handsome piece of machinery with complicated masses of gleaming steel and polished woodwork. It had the teething arrangement set in a revolving shutter like a small edition of an ordinary shop blind; and by automatic adjustment the crushed and shredded fibres passed through it for the last time in a long flattened stream of accurate and even thickness. This is termed " sliver," and it is coiled down into long metal cylinders known as " sliver-cans."

By this time the material has been cleaned and prepared. Some of it is still hard and fibrous, and will remain so, as may be ascertained by examining the ordinary three-yards-a-penny clothes-line. In comparison with its crude state, however, it is soft, and other varieties which go to the making of better-class articles are as soft as wool, some of it being of a beautiful, creamy nature, impossible to identify with the bales of raw material I had passed out in the yard.

The "sliver," still bearing but slight resemblance to rope or cord, is taken away in the cans to the twisting machines, and from there it is reeled off in one long thread on to bobbins. I am afraid I am not equal to describing with any clearness these twisting machines, and yet to see them in action they seem, with all their complications, simplicity itself.

"Ho, there! Mr. Artist—a twisting machine, if you please."

And now we have the groundwork, so to speak, of our rope, cord, or twine. In every case the process is the same from this time forth. It merely means more twisting and more strands—strand within strand in two or three ply, or more, and rope within rope until the required size is obtained.

The process is wonderfully expeditious and neat. The "sliver" makes its way to the various completing machines in the cans, and

BALLING UP TWINE.

the inevitable bobbins, which are everywhere,
carry the threads from process to process.
Three of them, to give an instance, will feed

their contents into a twisting-machine at one end, and one will take up the manufactured product at the other, to be joined later, in all probability, by two more, and spun through a machine on to another bobbin.

The rattle and roar of these machines is something indescribable, and my artistic friend is not sufficiently advanced in his profession, either, to convey an adequate impression of it, though he might contrive to catch the murky effect of the dust that tries the eyes and lungs of the unseasoned. It looked dangerous, too, in amongst those closely-grouped masses of revolving machinery ; but here, as everywhere, due precaution is taken for the safety of the *employés*. Every machine is fitted with a lever —with two generally, one on each side—and although the engine is never stopped, except in grave emergencies, until the mill closes down for the night, a pull or a blow shifts the belt on to a fly-wheel, which travels loosely on the shaft, and brings the machine to a standstill.

Out into the yard again, and not sorry to be there, fog notwithstanding, after the rattle and clash within those hives of industry, and I found myself looking wonderingly into a long

trough of boiling tar. It was, I discovered, a
tar bath, through which the strands of cordage
are passed before being twisted up, when
their outdoor uses and the action of sea-water
to which they are to be exposed render its
preservating qualities a desideratum. The
fishing-fleets are supplied with a large quantity
of this tarred rope for trawling, *i.e.*, towing
their nets, and a thousand and one other purposes, and this unprepossessing-looking bath,
small though it is, was in pretty constant
demand. I should be afraid to state how many
miles of rope pass through it in the course
of a twelvemonth, and I had asked so many
questions already that I had not the impudence
to go in for statistical inquiries of that sort.

The vegetable product known as Archangel
tar—uot the ordinary black article most of us
are familiar with that comes from coal—is
used, and is kept fluid by an arrangement of
steam-pipes running round the sides. The
yarns are uncoiled from an immense drum and
are immersed in the hot tar. An ingenious
arrangement meets them at the foot and
squeezes them more or less dry, according to
the requirements of the finished product.

The "walk," which runs the whole length of one side of the yard, is laid down with rails, with heavy rope cables down the centre of each. It is gas-lit, and the effect of the whirling machines travelling up and down through the fog like uncouth spiders, spinning out rope as they travelled, was weird. It takes two of these machines to manage a rope—one stationary and the other travelling. They are something like an ordinary gun-carriage in appearance—I am showing myself singularly apt at comparisons—and at the outset they stand together. The twisting arrangement on each revolves in a reverse direction, and the supplies are received from a stationary rack set just behind, to which the bobbins, each carrying one strand, are taken from the mills where we have seen them being prepared.

According to the size and nature of the rope to be made, so many of the strands are led away through the stationary machine, and hooked on to the traveller, which is then started off. Away it goes down the walk, twisting up as it travels, and leaving a gradually lengthening rope of complete make and finish in its trail.

Now, supposing something big and pretentious is required—a cable, we will say! Then, as in the case of making the strands, the pro-

PLAITING THE RUBBER CORE.

cess is repeated, only, in the place of strands or threads, ropes of quite respectable size and strength are attached to the machines and twisted up in the self-same way.

At the bottom of the yard, a number of hand-machines are still at work, for old customs die hard, and prejudice, where customers whose fathers had *their* ropes twisted up by hand are concerned, has to be respected.

Machine or hand, anything of the shape and nature of rope can here be supplied in any length and of any thickness, from a gigantic cable to packing cord and shop twine.

In the twine building, the same processes are in operation, saving and excepting the polishing and balling. Most of you must have observed the smooth, sleek, shining appearance of a piece of twine, and this is obtained by running the threads through a bath of size. This is a messy, though not a badly-smelling process. It is then led over and round an immense tin cylinder, which is heated to a regulated temperature by steam, and is finally submitted to a brushing on cocoanut fibre matting, which puts it in suitable trim for the balling machines upstairs. These machines are also ingenious in their way. They are tended by girls, whose duty it is to operate a lever which enables the twine to be coiled up evenly and neatly. Others, again, are engaged in finishing off, weighing and measuring.

"Balling machine, please, Mr. Artist." Handy fellows, these artists—save a lot of trouble!

The three or four loops round the centre and the fastening off still has to be done by hand, the mechanical appliance stopping at the actual coiling-up process.

A further branch has been fairly recently added to Rope-Land, although, strictly speaking, it has no concern with rope. This is the manufacture of steam-packing for cylinders.

It is a wonderfully pretty sight, albeit messy to the last degree, and smelly, too, with French chalk and boiling tallow, with the gleaming white threads running down from bobbins arranged vertically above the machine, which plaits it around the india-rubber core that is shown in the illustration running down over the guiding pulley-wheels.

"Service, please, my artistic friend!"

The india-rubber core is first prepared, and cut square and shapely by a band-knife, an endless band of thin steel which runs over a couple of pulley-wheels, one above and the other below.

I left this department with real regret, for to me, at least, these machines have an overpowering attraction, and if it were not for my energetic nature, and certain other claims on my time and attention, I could stand watching their ceaseless activity for days together, with, of course, a break now and again for rest and refreshment. However, I did tear myself away, and, after running over for one last view of that great giant, still thump, thumping away in the engine-house, as though time were naught, and the pigmy and sordid little considerations of everyday life were of no interest or moment—not worth three drops of the oil that a fussy little pump is continually pouring over its bearings, I thanked my guide, and went home to tea, lost in amazement—even as my train subsequently got lost in the fog —-and astonished admiration of the many marvels I had seen, heard, and gathered in Rope-Land.

IN TRAM-LAND.

TRAM-LAND is situated — cheerfully situated, I may say, seeing that it lies between a cemetery and a workhouse— in a quiet and somewhat mean thoroughfare, within easy walking distance of Leyton Station on the G.E.R. A more painstaking and conscientious explorer than myself would, perhaps, have started at the other end of this thoroughfare. Walking along in dreamy reflection and mud—the first is natural to an explorer on duty, and the last is natural to the Leytonstone Road—he would have observed a junction in the tram-rails that he could have followed down, had he been so minded, from Aldgate. A branch line here takes a sharp curve down a narrow turning, and is lost in a labyrinth of houses. This would doubtless interest the explorer and engage his attention. He would hitch his trousers still higher out of the mud, grasp his umbrella tighter, and would plunge in after that mysterious rail.

If he were gifted with a big imagination— and he wouldn't be much of a success as an explorer if he were not—he would, on coming to the Union before mentioned, assume that the line had been laid down for the convenience and comfort of aged and infirm paupers, so that they might be " delivered free at the door," like our groceries and railway parcels. It is miserable to have to walk to a workhouse, especially in inclement weather, and this instance of kindly forethought on the part of the Guardians would probably strike him as re-markable—it would me !—and cause him to further investigate. He would then discover that there is no tramroad to the workhouse, and that the East-end paupers, as elsewhere, are constrained to cabs or pedestrianism.

The rail still runs on past that home of wretchedness and woe, and, slowly and sorrowfully may be, the explorer resumes his travels. Still further along the road the lines take another detour, and are lost behind a pair of huge wooden gates, which bear the inscription, "North Metropolitan Tramways Co."

I knew all these things, and was spared as much mental exercise as an uninitiated explorer might have experienced, although previous acquaintance with the neighbourhood did not spare me from the mud, which was everywhere, nor any bit to eat, as the Ancient Mariner would doubtless have expressed it. I walked in from the other end, and was in Tram-Land within ten minutes of leaving the station.

Tram-Land is spread over four acres of land, and is composed for the most part of a series of long, lofty, well-lighted buildings, all looking marvellously clean and tidy, except the smith's shop, which is not of aggressively immaculate spotlessness.

Many of these buildings are used as store-rooms, some are fitted with bins and cupboards, where paints, bolts, screws, and all other fittings required for the making and working of trams are stocked. Others are for the storing of harness, of which one room alone held some-

thing like £3,000 worth, and for the fancy
woods, such as the prettily-grained bird's-eye
maple, for interior decoration.

Mr. Norris, the company's foreman, who
seemed to be pretty continuously engaged by
the oversight of such an extensive charge,
kindly spared time to take me round, and
point out to my inquiring eyes the manifold
marvels and mysteries of the land I was
exploring.

He was deep in an electric-lighting scheme
when I found him, it being his intention, I
understood, to wire the whole works. A
number of the shops are already fitted, and are
supplied by a small dynamo, which finds a
home in the spacious engine-house.

We went into that building first. What
struck me most — omitting the spring-door,
which came back on me before I had properly
got inside, and fetched me a slight refresher
over the knuckles — was the superior, conscientious attention to business details that
these engines evince. There are two of them
on opposite sides of the room, though only one
was running, and despite the fact that Mr.
Norris unlocked the door when we entered and

A CORNER OF THE ENGINE-ROOM.

locked it after him on leaving, and the total
absence, as far as I could discover, of anything

in the shape of an engineer, it went on steadily,
quietly, and unobtrusively supplying the motive-
power to the whole of the machines in the
works.

The timber for the heavy work is stacked
under cover in a long shed, where it is kept for
four years, so that it may be thoroughly
seasoned. The English and American varieties
of oak and ash are mainly in request, and huge
lengths, separated by cross-ties, are laid plank
for plank in the form of the tree they were cut
from, and cover many feet of what the auc-
tioneer would describe as " eligible building
land."

From this stock the timber is taken in proper
rotation to the mills to be variously cut up and
fashioned into something for horses to spend
their superfluous energy over.

My guide told me how tram-cars are constructed, and illustrated the process to the best
of his ability; but owing to the method of
dealing with the orders, my view of the operation was confined to observing the finishing
touches being put upon a series of fourteen
heavy timber platforms, which go to form the
flooring of fourteen respective cars. They

build them in sets of fourteen in these works, and start them and finish them off together, so that, no matter when the spectator visits the works, he will only be able to see one particular stage of the operations going forward, although he can see it fourteen times over if he likes to walk the whole length of the building.

One car a week all the year round is the average turn-out, I was informed; and, sure enough, later on, when I came to the finishing-shop, there was the last batch of cars, standing end to end, like a saloon train built on the American system.

I visited the saw-mills first, and saw the huge planks being cut up and shaped by machinery similar in nature to that which I described on my first exploration. Businesslike-looking circular and band saws were biting through whatsoever came before them in the way of business, in a bustling, impartial way that inspired respect and caution. "My business is wood," they might seem to observe; "but if you like to make it fingers I daresay I can accommodate you."

One noisy, clattering, and frisky little com-

bination of iron and steel very much engaged
my attention. It was a morticing machine, and
stamped a straight-edged tool up and down on
the material placed under it, biting out the
wood squarely and evenly as it travelled. I saw,
too, several fussy little rabbiting-machines at
work, artistically carving out the ledges for the
insertion of the windows in the side panellings.
They are simple and rather insignificant-
looking tools when they are at rest, standing
up vertically in the centre of their steel bed-
plates; but when they "get a gait on them,"
as the Americans would say, of something like
2,500 revolutions a minute, and begin to show
you what they can do, you are fascinated, and
could stand watching them all day; if they do
not affect you like that there is something
lacking in your constitution, and these papers of
mine will have, I am afraid, no interest for you.

From the sawing, rabbiting, planing, and
morticing machines, the various parts, fashioned
and finished, find their way—that is a figure of
speech, for of course they are carried in direct
—to the building I before mentioned, where
the tram is built up from a platform such as
the one I saw being constructed. It is a simple

IN THE SAW-MILL.

process though a mechanically perfect one, and I much regretted that I was not able to follow it out personally in detail.

Strength is, of course, a prime necessity, and none of the parts are allowed to pass with anything in the shape of a flaw in them. Rib for rib, cross-ties, panels, and the window-frames, which are made in the one piece to run the whole length of the car, the parts are brought in and adjusted until the body is completed in the rough.

In the meantime the wheels, "draw-bars," and other metal parts and fittings have been preparing, and in order to see the operations we have to make a tour through the engineers' shops and the smithy.

Good, solid, substantial sort of machines we find here, with no finnicky arrangements about them such as the wood-workers exhibit. They are built for plain, hard usage, though they handle the tough material they are employed upon as skilfully and adroitly as their prototypes do in the other departments.

Watch this big fellow at work planing down a bar of iron to a required measurement! To and fro, without cessation, goes the blade in calm, leisurely sweeps, that betoken neither haste nor impatience, but a placid, business-like devotion to the work in hand. In fact, if

any of these heavier machines have a vice at
all I should say it is that of bumptiousness—it
is too serious for conceit. Conceit argues a
certain amount of levity, and is apparent along
with a *soupçon* of vanity in the more giddy and
unstable wood-workers. But you don't find
anything of the sort in the engineers' shop.
They there take life and its monotonous duties
very seriously indeed. The planing machine,
in particular, I should say, has been the victim
of some misplaced attachment early in its
youth, and has gradually acquired a morbid
pessimism that leaves it nothing to hope for
from the future save iron. When under the
influence of an unusually generous supply of
oil, it becomes mellow and forbears to groan at
its labour; ordinarily it sighs and creaks dismally, as though it would say: "Ah! woe is
me, but that's a tough 'un." One of these
days it will drift into a state of melancholia,
unless it turns to Socialism. And yet, if it
were not for the crushing effects of a disappointment in its early youth, I fancy it would
have been rather cheerful than otherwise—not
exactly a humourist, but given to irony of a
ponderous and rather a biting nature.

"Ah!" it would then banteringly observe to
the engineer, on having a six-inch block of
metal placed under it, "why these delicacies?
Why not bring me something stout and solid—
something I should have to tussle with? Bring
me the heart of a man who promotes building-
society fizzles, the conscience of a money-
lender, the brain of a magistrate off the pro-
vincial bench, the cheek of a journalist, or the
delicacy of a *Star* reviewer, and take this trifle
away against the time when I am feeling
languid and need a rest."

It is a pity that it is such a heavy-minded
machine, for it would have many an opportunity of throwing off a brilliant *jeu d'esprit*
along with the metal shavings in the dim re-
ligious light of the corner wherein it stands.

The wheels and bigger fittings are received
at the works in castings, and are turned up,
and bored out, and made pretty and useful by
one or another of the massive machines that
stand around me. Other parts—such as the
axles and "draw bars"—are forged out of the
rough bars in the smithy and brought in here
to be finished off afterwards.

In the dim gloaming of the December after-

IN THE CAR-BUILDING ROOM.

noon that I visited Tram-Land, the smith's shop
presented a very weird appearance with the

forge fires dotted about it, and gleaming bars
of metal throwing off a brilliant rain of red fire
under the vigorous poundings of a couple of
steam-hammers.

It is astonishing how soft and pliant the
heated metal seems to become under those
hammers. It is crushed out, and rounded and
flattened up again, bent this way and that,
patted here and patted there, and lastly is laid
between a mould that opens in the centre to
receive it, and fairly squeezed into the desired
shape just as simply and unconcernedly as the
butterman manufactures his pretty pats of
" best Dorset."

The forges here are all driven by steam fans,
and the roar of the fires, the clanging of heavy
hammers, and the rattle of machinery, along
with the steady pound-pounding of the big
steam-hammers, leave an impression that must
be imagined, for I cannot describe it.

Both shops exhibit a number of ingenious
labour-saving devices, including one for shrink-
ing the tires on wooden wheels. Wheel and
tire together, shrunken and expanded respec-
tively to their greatest extent, are run on to a
flat platform, a catch is released, and the whole

affair sinks into a huge basin filled with water. Another contrivance worthy of note is that for cutting up cold metal. A pair of powerful shears set at an angle close on the material which is placed between them slowly and with sinister intent, as though they were rather gloating over the job, and with a groan and a wrench bite their way through. A two-inch bar of tough iron was chewed up for my especial delectation, and I marvelled thereat, and put my hands in my pockets. Amputation is such a simple and suggestive operation when near these machines, and I had further use for my fingers.

Turning away from these mechanical attractions, we entered the harness-making shops, where traces, collars, and other horse impedimentia were being variously constructed. Most of the work here is still done by hand, for though machinery is provided, nothing stands the hard strain like the good, old-fashioned wax-end stitching. One very ingenious contrivance was shown me for sewing up sacks, binding the two edges of the coarse material together with a stout, tarred cord, and making quite an artistic job of these future repositories of oats and grain.

From this building we came out upon the
last batch of completed trams being painted
and finished off ready for the road. Steam-
driven mills grind and mix up the paint and
turn it out by the gallon, so that there is
nothing left to do but to put it on. It is an
amiable and engaging composition is paint, I
was told, and it always leaves a good impression behind ; but I didn't go and stand up
against it to see.

Somewhere about here I came across a
further instance of the kindly consideration
of the general manager, Mr. R. L. Adamson,
for the comfort and convenience of the *employés*
in the shape of a well lighted and arranged
mess-room, in which the men have their meals
cooked and served.

In addition to making trams for their own
line, I learnt that the company had numerous
outside contracts from all parts of the world,
and in consequence there is no slack time in
Tram-Land. If there is any cessation in actual
construction there are always a number of cars
needing repairs, alteration, or decoration, to
say nothing of the trifling matter of a few
hundred 'buses and waggons and minor rolling-

stock which the company has to maintain, and consequently to build.

I had a few minutes' chat with Mr. Norris

MOUNTING CARS ON WHEELS.

in his spacious office anent Tram-Land and trams generally, as I prepared to leave.

In speaking of the future of the tramway companies, I suggested that the County Council might possibly elect to adopt a policy of amalgamation.

Mr. Norris looked oddly at me. "Ah, I see!" I observed. "It would be the sort of amalgamation that takes place when the cat is left alone with the canary; perhaps not quite satisfactory for the canary. Well, 1 won't detain you if you have nothing more to show me."

"I think not," he replied.

And so I left him, and took the rail again, and was shortly after come unto Battersea, when—— but that is another story.

IN CANDLE-LAND.

NATURE was in a facetious mood the day I went to Candle-Land.

"Here's a lark!" she said to herself, with a chuckle. "Andom's going out exploring a-gain. I'll turn on a little weather for him "— and she did.

First it rained a bit, just to get me interested; then it fogged, then hailed, then thundered, while the lightning struck the church steeples, and the bitter north-east wind roused itself from its lair in the north-east regions, and swept with a dreary howl across vast continents and icy plains, spreading desolation in its path. Half-way down the Victoria

Embankment it met me and stopped for a little friendly conversation. I was busy picking my hat out of a mud-heap at the time, and couldn't attend to it then, so it went on to Battersea and waited for me there.

I went down by a slower route, *viá* the L.C. and Dover Railway. Being charitably inclined, I thought I would give that line a turn. It looked as though it wanted a little patronage. The porters seemed surprised to see me. They were grateful, too, in their sturdy, independent fashion.

"Here's a chap wanting to go Battersea way," they seemed to observe. "Got anything going there?"

After a little consultation, they decided that they had, and they found me a passenger-truck and put me in it with a flattering display of incivility, and left me to my fate.

When I got to Battersea and found that I was still a mile and a half from my destination, I took myself severely to task for my unseasonable generosity. Fourpence is a good, large sum to spend on a deserving object all at once; but it doesn't go far in dirt and incivility when you have such a large field to cover, and the fractional benefit I had conferred was at the expense of my comfort and well-being.

However, I found the north-east wind and a mud-heap waiting for me outside the station, and I and the former walked over to Candle-Land together. Arrived there, I parted with my companion for the time being, and went inside to explore its many marvels and mysteries.

The recollection of my visit to Price's Patent Candle Company's works, to be hereinafter known and designated as " Candle-Land," will always be a pleasant one, both on account of the extent of the firm's operations and the interest I had in them, and by reason of the courtesy and kindly consideration that was accorded me. Vanity is the besetting sin of mankind, and even an exploring scribe doesn't mind being mistaken for a gentleman now and again, and treated as such. It gives him a fictitious belief in himself that is soothing and vastly pleasant, all the musty old rubbish that can be advanced to the contrary notwithstanding.

Mr. McArthur was deputed to take charge of me and direct my explorations, and together

we strolled out of the offices, across an open
yard littered with hundreds of casks and
barrels; and the wind kept us company the
while.

"Now," said my guide, pleasantly, "what
shall I show you?"

"Everything," I replied, promptly.

He laughed, and led the way across a light
ironwork bridge connecting the works that are
spread out on either side of a creek, conveniently running up from the river a few
hundred yards below. Up this creek the
company's own lighters and barges bring
annually thousands of tons of raw material,
and bear away the finished product in
quantities correspondingly large.

Candle-Land displays in its outward aspect
a bewildering succession of buildings which are
for the most part of one-story construction,
long and lofty, with arched roofs of galvanised
iron, through which skylights admit a workable sufficiency of the light that cheers and does
not inebriate. Topographical and statistical
information is hardly called for; neither do I
feel altogether impelled to dilate on the many
patents and processes by which Price's candles

have been contrived, constructed, and improved. All these matters are important, and compel admiration; but they hardly come within the range of my pen, and in the interests of more weighty subjects I must pass them by.

Candles have their beginning in a barrel of fat. My explorations came perilously near to resembling candles in this respect, by the way, owing to my own clumsiness and the slippery flooring. I managed to save myself at .the edge of a simmering tank of the material, however, and no damage was done; but I did not altogether contemplate becoming a " dip," and I was extra careful thereafter.

Technical gentlemen like my guide call it " paraffin scale," " palm oil," and sundry other high-class, important-sounding names; but I called it " fat," though, as a matter of fact, the opening of a barrel of paraffin scale revealed to me a yellowish dry and powdery substance which is extracted mainly from American petroleum.

The heads of the barrels containing this material are knocked away and the contents are shot into immense underground tanks capable of holding six tons, where they are reduced to a liquid state by steam-heated pipes.

One has to imagine a great deal when viewing these operations, because, unfortunately, they are carried on in regions not readily accessible, and I did not conceive that duty imperatively called upon me to go a-swimming in order to give a graphic description of the huge tank of simmering "scale" lying below me. Even if I had, I fancy duty would have had to call pretty loudly to make me hear.

When melted, the paraffin is pumped up into a series of overhead tanks, and by this time having deposited oddments in the shape of any foreign substances that it may have accumulated, it flows through pipes into a number of shallow metal trays, stacked up in an adjoining chamber on a series of horizontal platerack affairs. Pipes traverse this room in all directions, and the workmen fill the trays by means of a flexible attachment with a broad flattened nozzle in very much the same manner that the peaceful suburbanite waters his geraniums and other children of nature out in his own back garden.

In these trays the paraffin crystallises into

MIXING THE MATERIAL.

portable slabs, which are then conveyed to
heated closets and subjected to a baking in a

carefully-regulated temperature. Under this
gentle moral suasion the oil and softer material
flows away. The "base," so to speak, is once
more rendered liquid, and stirred up in a huge
metal cylinder termed an "agitator," along
with a supply of animal charcoal. As the
result of this treatment a beautifully clear,
hard, and white substance is obtained, out of
which the ordinary candle of commerce could
be—but is not—manufactured.

To make a really good and satisfactory
candle another material is required—later I
will tell you why—and this requirement necessitates in its turn various complex and delicate
operations before the workable commodity can
be got at. "Stearin" is its name, and either
palm-oil or tallow may stand to it in the rela-
tion of parent. In plain English, these fats
are compound substances which yield on de-
composition fatty acids and glycerine or "sweet
waters"—a beautifully-phrased description
which honesty compels me to state belongs of
right to Mr. McArthur.

To obtain this desirable end the fat is mixed
in true and just proportions with water and
lime, and is heated for five hours in a large

copper boiler termed an "autoclave," under a steam pressure of 120 lbs. to the square inch. This has the effect of decomposing it—something much less elaborate would serve me if I were yearning to be decomposed: but these whims have to be respected—and a mixture of fatty acids and lime-soap results, while the glycerine remains in the water, to be afterwards recovered and purified. At least Mr. McArthur assured me it does, and he ought to know. After the lime has been extracted from the resulting product, a treatment with strong sulphuric acid at a high temperature, and a washing further purifies it. By this time it is black in colour, and not at all tempting in appearance. But they know how to manage these matters in Candle-Land, which is fortunate for the public at large. I am sure I shouldn't have the vaguest idea how to proceed. I should get frightened at what I had already done, in fact, and sneak out of the back door before I could be arrested, and go and join bogus company directors in the land of liberty beyond the seas. Not so these people!

"It is a bit dirty," they remark contemplatively, and they turn it into another huge

copper contrivance, where it is distilled by
super-heated steam, and finally condense it in
a series of vertical pipes, which look vastly
imposing and important as the illustration
shows.

Even then the treatment is not complete
until the fatty acids, crystallised once again
into slabs, have been put into canvas bags and
placed in powerful hydraulic rams, where they
are subjected to hot and cold pressure for the
removal of the oleic acid. I don't know precisely what oleic acid is; but it is a liquid body,
and I presume they don't want its services
there, or they wouldn't be at such pains to get
rid of it. And thus the stearin, hard, white
and opaque, is obtained for the candle-maker.

There are many features that I have passed
by from sheer inability to grasp them in a
presentable form, and there are others that considerations of space prevent me from dilating
upon. Wandering through the rooms in which
the various processes are being carried out,
huge vats and tanks of melted material strike
the eye, even as smaller quantities splash over
and stick to the clothing. One does not need
to put on his Sunday-go-to-meeting suit to

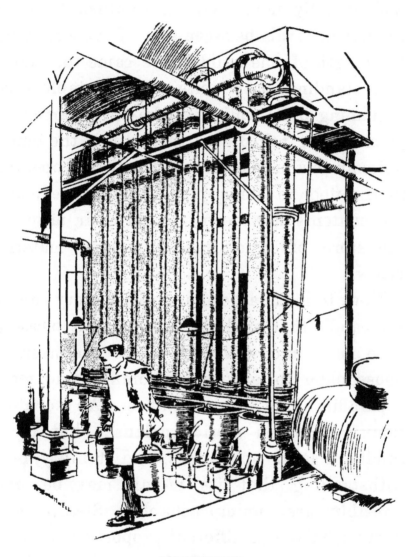

CONDENSING.

go exploring in Candle-Land. Storage and
melting tanks are everywhere, though I never
imagined, as I walked over the iron flooring

and casually noted several smallish holes at
regular intervals between sets of rails running
the length of one building we came to, that I
was standing over quite a respectable-sized
swimming-bath, full of palm-oil. Barrels of
palm-oil are rolled over these lines, the bung-
holes are made to correspond with the opening
in the floor, a steam-pipe is introduced, and
the contents flow down to add their mites to
the store of crude candle - making material
below.

Now, to make a good and satisfactory candle
—a candle that shall be a joy and a blessing
unto man, and shall not cause the average
householder to lapse into profanity and tears
—paraffin alone does not suffice. As a contor-
tionist, it would merit applause; but for a
candle, it lacks uprightness and stability, notwithstanding that its morals may be, and
probably are, unimpeachable. Stearin and
paraffin, mixed in different proportions according to the quality of the candle desired,
obviate this difficulty, and a considerable
quantity of candles are made of stearin alone,
which must be a humiliating reflection for its
companion, if it is at all of a sensitive nature.

In case I did not mention it, and supposing that you are yearning to know, the liquid paraffin, after all the processes of melting and cleaning it has undergone, is pumped away through filters into huge iron tanks, from which it is dipped out into smaller wooden vats mounted on wheels, and run through to the various departments by horse-power.

In a spacious building containing many more wooden vats of gigantic structure, the paraffin and stearin are practically introduced to each other, and so thoroughly do they assimilate that it is difficult to discover which is which. "Which was which, I couldn't make out, despite my best endeavour," as the song runs; so I merely noted that each vat is hypothecated to one especial kind of candle, and turned aside to follow the actual and more interesting process of candle-making. All this time I had been occupied with the preliminary operations of mixing and preparing the material.

"Dips" still survive, and in a comparatively small room I found the process in active operation. As the name implies, "dip" candles are built up by a succession of immer-

sions in tanks of melted material. The wicks
are first of all wound upon light iron frames,
each holding sufficient for 100 or 120 candles—
vide Mr. John Calderwood's excellently clear
little manual on the process of candle manufacture. My friend Whitwell's artistic efforts
here will doubtless assist to make the operation
clear to you. They dip, and dip, and dip, and
still they go on dipping, until the pendent
candles counterbalance the weights contained
in the tray just under the pulley-wheels, when
they cry—the operatives cry, not the weights—
"Hold, enough!" because they—the candles—
do hold enough.

It may occur to the untrained mind that this
process must be very slow. In fact, I myself
couldn't see why each successive dip into the
material didn't melt off the previous coat, and
so on until after a year of hard labour the
workman found himself gazing ruefully at a
string of wicks and wondering how they came
there. Mr. McArthur explained to me that
the temperature of the material was too low
for this to happen, and the explanation satisfied
as well as pleased me by its very simplicity.

In the candle-moulding room, the scene is

much more varied and striking, and I know no
other example which so strikingly contrasts the
"old style with the new." Here they turn out

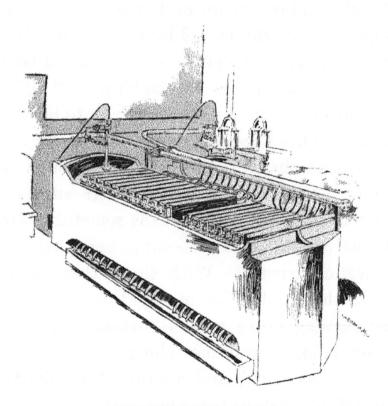

THE HYDRAULIC PRESS.

candles by the thousands by a brief preparation
aud the lifting of a lever. The illustration will
give you an idea of the nature of a moulding
machine. The wicks are contained on bobbins

placed under the machine, each separate wick
running up through a small iron cylinder,
which is moulded on the top in the familiar
candle-top shape. A crank actuates this system
of rods, and by the upward pressure pushes the
candles out of the mould into a pair of adjustable clamps, in which they are removed bodily
from the machine. In making a "fill," a
current of steam is first turned on to heat the
moulds to the necessary temperature, and then,
from a number of vats stationed round the
walls, the operatives obtain their supply of
material, dipping it out in cans something after
the nature of zinc hot-water cans used for
domestic purposes. With these they proceed
to fill the mould, the material flowing along,
and settling down until it overflows into the
grooved basin running along the top. The
steam is shut off, and another tap admits a
current of cold water, which circulates about
the mould until the candles are cold. The last
completed batch of candles is still standing in
the clamp-frames above the machine, and is
left there until this stage of the proceedings,
when a sharp knife severs the wicks between
the two sets, the original batch is emptied out

on to a receiving table, the clamps are restored
to their place, and, after removing the overflow
of material with a putty-knife sort of contrivance, the workman turns the crank, and a
further contribution of 100 candles emerges
into the air. They in their turn remain in the
clamps above the machine, partly to solidify,
and partly to hold the wick in its place while
the mould is again charged, and so the process
goes on *ad infinitum*. The " self-fitting " candle,
a speciality of the firm, and the invention, I
believe, of their Mr. George Spicer, differs
slightly in moulding from the preceding operation. The fluted conical butt is obtained by
dropping a number of small, removable moulds
into enlarged spaces at the top of the candle-
moulds made to receive them. The material is
then poured in and " fixed " and turned up as
in the previous process, the candles carrying
the moulds up with them as they emerge. A
slight tap on the end serves to release these,
when they are transferred with wonderful
rapidity to a number of pegs ranged on a board
just over the machine, ready for the next
filling.

Leaving the moulding-room, filled with an

overwhelming sense of the supremacy of man and his marvellous handiwork, I tripped over an empty packing-case and limped into a building in the near vicinity, where tapers and night-lights were being variously constructed. Night-lights are an important feature in the industries of Candle-Land, and many hundreds of thousands of boxes are turned out annually. They are either moulded and turned up much in the same way that candles are manufactured, or they are filled by hand into little cardboard cases which are previously wicked—not wicked in the sinful sense—and prepared.

The energy and despatch of the hand-labourers in these works is something to marvel at. Take the operation of filling the night-light cases, for instance. The cases are set out on a table in consecutive rows, and to them will come a young fellow bearing a small can of creamy preparation. Then by a series of jerks the contents of that can are emptied into the cases, and never a drop finds its way over the sides or on to the table, which is marvellous. I could cover the table and fill my boots, and squander it down my wardrobe easily enough; but I would not engage to get it into

"DIP" MAKING.

the cases. I know this because in a moment
of temporary aberration I once tried to fill a

6

paraffin lamp, and—but that, as Mr. Kipling
would observe, is another story.

The moulded night-lights, unlike candles,
are made without wicks—a spike taking its
place in the mould, and leaving a hole for the
insertion of the wicks in a subsequent operation. Handwork of marvellous agility is again
encountered in viewing the process by which
the wicks, previously attached to the little tin-
star at the bottom, are thrust through the
opening left for the purpose, before the night-
light goes on to join its companions in the
boxing and packing rooms.

Right at the back of the night-light-filling
building stand a pair of huge drums, which coil
up thousands of yards of taper. This is a
wonderfully self-contained process. The cotton
strands are placed in skeins on one side of the
building, and the drum revolving on the other
draws it through and through a wax-bath and
presents it at the other end ready to be cut into
lengths and bundled up for the market at home
and abroad.

A peculiarity of Candle-Land worth noting
is its independence in the matter of outside
trade assistance. It has its own saw-mills; it

makes its own machines ; its boxes are machine-manufactured by the thousand on the premises ; it does its own printing, and scores of other minor operations that one would not expect to find associated with the manufacture of candles.

On coming away from the box-making department, I turned into a building at the instigation of my guide, and was greeted as the door opened by an indescribable clatter and clash. It was here, I found, that the wick is prepared in a number of plaiting machines that present a very pretty sight to gaze upon, though one doesn't feel encouraged to attempt a conversation in their immediate neighbourhood.

"That is a pretty movement," I yelled to my guide, indicating a vertical machine whose delicate, complex arrangement aroused my enthusiasm.

"Yes, all cotton," Mr. McArthur howled back.

I tried again.

"It seems almost alive," I screamed.

"About a quarter past, I think," he yelled in reply, pulling out his watch and showing it to me. It was a quarter past five, had I been

wanting to know; but I concluded to accept
the information, and postpone comment until
we got outside again.

They do say that old stagers, men who have
grown up in these mills, beguile the tedium of
their labours by telling each other fairy stories,
and by reading Ruskin and other improving
masters aloud to each other.

Strange, passing strange! but it just shows
you how familiarity set impossibilities at
defiance.

More mechanical marvels were brought to
me, speaking figuratively, when we got outside
and were placed in front of me, and I wagged
my umbrella and ejaculated in exceeding
delight and astonishment. And then I came
to the last scene of all, the inland and export
packing and dispatching buildings. The pack-
ing-rooms display further remarkable evidence
of the dexterity and precision in hand-work
that can be acquired. The candles are made
up in one-pound paper packets for export, and
in three and six pound packets for home use.
The packers, from long·use, will grasp a
handful of candles from a box by their side,
will reject three or four, or perhaps only one

CANDLE MOULDING.

from that handful, and in "one, two," a six-
pound package is being sealed with wax by a

boy at the further end of the counter. It is by the sense of touch only that they can possibly know that those packages are correct in weight and contain the regulated number of candles. An ordinarily expert packer can make up something like 3,000 packets in a day's labour, and he does it, too.

In affixing the company's labels to the wrappers, which is done by quite young boys, the same smartness is observable. The process seems to involve a deal of wriggling. I thought they had all got St. Vitus's dance; but it seems that it is a customary habit for them to put themselves into violent motion, probably in order to obtain sufficient momentum to overcome the resistance of 5,000 wrappers per diem, which is the average reckoning of an expert.

In the packing department also, the candles for export are packed in small wooden cases which are addressed by a very rapid stencilling process, and eventually, I believe, find their way back to the wharf where they originally arrived in a very elementary condition, to, perhaps, the very barge that brought them up, and so, as Pepys would have said, abroad.

Candles, and perhaps night-lights, which are in a sense miniature candles, are, appropriately, the product of Candle-Land; but just by way of relaxation and amusement the company turn out a few hundred tons of soap now and again

FILLING NIGHT LIGHTS.

to keep their hand in. As I had formulated a plan to explore an independent Soap-Land at some future date, I did not deem it considerate to trespass wantonly on my guide's time and attention. Another industry that, carried to a successful issue, has given the firm an im-

portánt standing on its account alone, is the
manufacture of glycerine, which I casually
referred to in its elementary stages, in describing the extraction of stearin.

And so, pleased and impressed by what I had
seen, I recrossed the creek, and turning up my
collar, and casing my hands in my pockets
against the friendly, though chilly, overtures
of the north-east wind, which had borne us
company wherever it could get at us, and had
evidently sworn never to forsake me, I thanked
my guide for so handsomely allowing me to
victimise him, and quitting Candle-Land by
the back door, made a bee-line for the railway
station, and was shortly afterwards come unto
the suburbs, where——. But my candle has .
been burning long enough, and so, in mercy
to my readers, and more particularly to myself,
I will—snuff it.

IN GAS-LAND.

FINDING my candle so unceremoniously snuffed, I immediately set about discovering another illuminant, and I found one at the West Ham Gas Works, which are Stratford way.

At the end of a quiet little street turning out of the main road, just past Stratford Market Station on the Great Eastern line, I discovered it, and it struck me as being uncompromisingly ugly. Nothing here for the painter to linger over, I reflected, neither would a poet be likely to go into rhapsodies over the scene. However, I was not searching for the picturesque. Rather did I hanker after the practical, so I made my way round to the secretary's office, and found Mr. Snelgrove waiting for me in a room rather more imposing than snug, though for all its size it struck pleasantly warm after the biting wind outside.

While chatting pleasantly over industries generally, and the gas industry in particular, Mr. Wright, the assistant-engineer, came in, and I was introduced to my guide through this particular land.

So far I have had the advantage of the common or garden explorer in being able to obtain a separate and qualified guide to every land I have passed through, which shows the advantage of doing your exploring at home. Had I gone abroad, I expect one greasy native would have had to serve my purpose for the whole series. Out in the open again, and we walked straight through Gas-Land to hunt up the initial stage in the process of changing coal at about sevenpence per hundredweight into gas at three shillings per foot. I used to make gas when I was a youngster, and I daresay many of you have done so too. It is a delectable and messy operation, and dear on that account to the heart of the average schoolboy. I did it

with a churchwarden pipe, a few ounces of
coal-dust, and a lump of clay, when the girl
had gone out and left the kitchen range un-
protected. The illuminant was not blinding in
its intensity; but the satisfaction that that
stewing clay and coal-dust gave me was beyond
expression. They have improved on my old
method in Gas-Land I observed.

I commenced my observations at the top
of a perilous-looking light ironwork ladder
some thirty feet above the level of the ground.
Mr. Wright scaled it with ease and the agility
born of use and long practice. I lacking the
ease, and being somewhat hampered by a
heavy ulster, an umbrella, a notebook, and one
lead pencil, arrived many minutes later, somewhat shaky, and out of breath.

This ladder, I may mention, is situated at
the end of one of the "retort"-houses, and
the temperature of a "retort"-house in the
depth of winter would be favourable for foster-
ing tropical plants, I should judge. When I
stepped off the ladder on to a staging outside
that runs round the building I found it fresh,
to say the least of it, though the view was
worth the climb and the cold.

Far below ran the canal, just then frozen and snowed over into a treacherous semblance of a solid supporting body, up which barges bring the coal supplies in forty-ton loads. Gas-Land lay stretched out around me, with tall chimneys, low, squatty buildings, and huge gasometers dotted about at irregular intervals.

Rails lead away from this staging to the various retort-houses, and the coal, landed by steam-cranes, is conveyed in metal cars and distributed about the works by quite a miniature railway service. It finds the ground-level by a series of shoots, which deposit it within convenient reach of the retorts.

I turned into the building again, and gazed down on the busy scene below me. It was an impressive and a striking sight.

A long and comparatively narrow passage leads from the open doorway at one end to the ladder above which I was standing at the other, between the coal supplies that are being continually replenished from above by the shoots, and the "retorts" surmounted by the complicated system of pipes. In reality the "retorts" divide one huge building into two, being built down the middle and so arranged that they may

IN A RETORT HOUSE.

be charged at both ends. They are set side by side, as the illustration shows, and are kept at a red-heat by immense coke furnaces. underneath.

I was asked if I would like to see into one of these furnaces, and I rather thought I should, until I casually discovered that the heat is liable to scorch the toes off one's shoes, to say nothing of doing superfluous and vexing damage to one's wardrobe accessories. I said, then, that it would be a pity to put the men to such needless trouble, and sternly negatived the protestations that ensued.

I learnt, however, that these furnaces never go out. Day and night, from Monday morning to Saturday night, the "retorts" are charged, and the process of making gas goes forward. On Sunday they cease and rest, but the furnace underneath burns on just the same.

A "retort" is simply a long, bricked chamber capable of carbonising five or six hundredweight of coal at one filling. The only opening from it, with the exception of the curved, oval doors through which it is charged, is the one leading to the pipe shown running up just above the mouth.

I wondered vaguely, as I leant over that rail and watched the men at work below me, why it was called a "retort." Later, the answer came to me, and I discovered that it was so quick at repartee that no other name could be given it.

"There's coal for you," the men observe, as they ram three or four hundredweight into its capacious interior.

"And there's gas for you," it retorts, with a snap and a burst of flame, as the men hastily retire out of its reach. Then one will run forward and slam the door to and screw it up, and you can hear it chuckling away to itself inside there out of the cold long after the joke has died from the remembrance of the busy throng, who have heard it so often that it has probably become irritating and stale.

These "retorts" are charged by means of iron scoops fourteen feet in length, something like a Canadian canoe in appearance, supposing the ends to be cut off square and a T handle fitted at the top. The bowl measures ten feet, and has a carrying capacity of one hundredweight, Mr. Wright informed me, though I should never have thought it possible. Three

times this scoop is filled and emptied into the
retort, one man being stationed at the handle,
and two others supporting the body by means
of a curved iron arrangement termed a " bridle."
This has a handle on each side; the bottom of
the scoop fits into the bend, and so the load is
borne into the " retort," and run home, the
men at the bridle hastily retiring to avoid the
rush of flame that takes place as the scoop is
turned over and withdrawn upside down by the
man at the handle. The work is done with
marvellous energy and quickness, because delay
here means loss of gas, and when the last fill
is made the contents are pushed back from the
entrance, the door is rammed to and fastened,
and the " charge " is left baking for six hours.

Each retort carbonises about one ton of coals
per diem; there are about sixty-four " retorts "
in one building, set in two rows side by side
and back to back. Gas-Land contains five or
six similar buildings, and with 365 days in a
year and coals at, say, twelve shillings per ton,
it occurred to me that I shouldn't like to foot
the coal bill. The same idea will probably
occur to you.

At the end of the six hours, the doors are

opened and the residue is seen running back in a long red line that is too trying to the eye to gaze in upon.

It is better not to be too eager to get a near sight at this stage of the operations. The

CHARGING A RETORT.

"retorts" have a spiteful habit of blowing out a huge sheet of flame and gas and dense sulphurous smoke when the doors are first opened, and if you are not expecting it the surprise is apt to be unpleasant and painful. The coke is

7

then drawn out, by means of long iron rakes,
into metal barrows, where it is cooled down by
the application of cold water, and wheeled out
to an immense heap, around which a whole
regiment of grandmothers could be instructed
in the art of mastication, publicly, instead of
getting surreptitious and imperfect lessons in
their coal-cellars at home.

I used often to be advised to teach my grand-
mother to eat coke, in my younger days, and,
assuming that it is a knowledge all respectable and conscientious grandmothers ought to
possess, it struck me that this would have been
an excellent opportunity to give the lesson.
There was such a sufficiency of material,
that the company could easily have afforded
the amount an innocent old lady of limited
appetite would have consumed in "swatting"
up the subject.

Now, we have seen the coal go into the
"retorts," and we have seen the coke leave
them—at least, I have, and you are seeing the
process through me—hardly so clear and impressive, I'll admit; but lucid and very improving.

In the process it has parted with several

properties, notably gas, tar, and ammoniacal water. All these things are comprised in the thick, yellow, smoky gas that rises up from the "retorts" through the series of huge iron pipes that tower in front of them. Arrived at the top, it turns over into an immense iron pipe termed the "hydraulic main," through which it is drawn by a set of powerful steam pumps or exhausters that are working in another part of the grounds. In turning, it deposits a certain proportion of tar which is drained away into tanks, preparatory to being emptied into an immense well, handily contrived out of the basin of an old and long-since discarded gasometer.

Even after it has rid itself of a few superfluous gallons of tar, the gas is hardly in a condition to be supplied to long-suffering householders, and the outside authorities would probably have something to say on the subject, even supposing the householder didn't kick—an unlikely contingency. So the crude gas is drawn away by the engines through the condensers, a series of upright iron pipes, which reduce the temperature considerably, and prepare it for the "scrubbers" and "purifiers."

The ordinary gas of commerce cannot be seen. It may be smelt, and a lighted match in the hand of a double-distilled idiot will discover its presence and make its whereabouts known to the whole neighbourhood. But up to the time of reaching the "scrubbers" it can be seen by even a near-sighted person. I cannot say that I myself did actually see it; but Mr. Wright assured me that such was the case, and I always take these facts from my guides, occasionally passing them off as my own.

Forced by the powerful pumps, the gas then goes on a circumlocutory tour, in the course of which it enters a " washer "—a huge horizontal contrivance half filled with water, in which revolve a number of wheels containing bundles of wood somewhat similar to the familiar half-penny bundles of commerce that are sold by oilmen. The wheels carry these wood bundles alternately through the gas and through the water, and as ammonia has the property of sticking like glue to a wet surface the result is obvious.

"So sorry," it observes hastily to the gas when it sees the wet bundle coming round in

THE PUMPS.

its direction, "but here's a friend of mine, and
I must go. Remember me kindly to all at

home. Ta-ta! I'll probably see you at the club to-night."

But it doesn't. The water in the tank below has too powerful an attraction for it; and once in, it stays, while the bundle, having disposed of that little lot, goes out to look for more giddy little ammoniacal stragglers. It suggests a conscientious policeman on duty—supposing there were such a thing—only that it is too intelligent, and it will not lie still, while the average policeman lies anyhow with a magnificent impartiality.

It takes ten gallons of water to wash the gas product of a ton of coal, to say nothing of the scrubbing process that it has to undergo in what are appropriately termed the "scrubbers." These "scrubbers" consist of two immense iron towers, which stand forty feet high and are filled with layers of coke arranged in tiers two and a half feet thick. At the top, ammoniacal liquor is poured down, and by the time the gas has forced its way through this tortuous length of coke and water it has left behind, in the form of ammoniacal liquid, all the ammonia it contained. After that it would seem superfluous to talk about purifying it. But they *do* purify it! These industrial people are so thorough and so wonderfully painstaking!

If it were left to the gas, I have no doubt that it would elect to go straight off and help twist round the wheels of some inoffensive citizen's domestic meter—perhaps mine. It is a light-hearted, frolicsome sort of substance is gas, and it delights in a joke of that sort, even though it cannot stay to witness the *dénouement*.

"He, he!" it chuckles, in a silly, irritating sort of way. "Won't he just say things when the bill comes in!" And it gives the meter an extra shove and goes rejoicing on its way to its one especial tap just under the baby's food-warmer.

I am glad they purify it, because, from what I know of gas, I am very sure it has a conscientious objection to the process.

This is done in a series of squatty, dome-roofed iron tanks, containing layers of lime six inches deep, which are termed with sweet simplicity and directness "purifiers." These number ten, and the gas takes them in regular rotation, entering at number one and leaving again at number ten, and then, for the first

time, it is gas pure, simple, unadulterated and
unalloyed.

The ammoniacal waters used at one time to
be a waste and an obnoxious product, for a
certain proportion of the ammonia being blown
off into the air, got the neighbourhood of Gas-
Land a bad name.   Being of a sensitive nature,
this grieved the officials, and they took and
collected it—the ammonia, I mean, not the
bad name — and sat down before it and
pondered deeply and darkly.

The outcome of their ponderings resulted in
a comparatively new building, where the
ammoniacal water is distilled and treated with
lime and sulphuric acid, and finally deposited
in the form of a crystallised salt known as
sulphate of ammonia.   This sells for an encouraging sum for use as manure on tea and
coffee plantations, and the officials have reason
to pride themselves on the results of their
stupendous mental effort.

Freed from the retarding influence of all
impurities, back the gas rushes to the meter-
house, anxious to get the remaining formalities
over, in order that it may let its light shine
before all men and increase the quarterly bill

WASHERS AND SCRUBBERS.

of one man in particular. Sometimes it happens that it may be disappointed at the last gasp, for it has many roads to traverse, and should it mistake the turning it might possibly find itself up a lamp-post, when the local authorities would have to foot the bill, and a pound or two more or less doesn't matter when the rates are eight shillings in the pound and the parish is " improving."

I followed it into the meter-house, which is a handsome, lofty building, gleaming with polished metal-work and filled with mysterious appliances, wheels, and fittings.

" By the way," I remarked to Mr. Wright, as we entered, " what do you make here ? "

" Why, gas," he replied.

" Oh, that accounts for it then," I rejoined, " I thought I could smell it."

Up to then I had neither seen nor smelt the product of so much pains and labour, and I had had to take my guide's statements on trust. This I was very ready to do, of course; but here was tangible proof, and it interested me accordingly.

The meters, of which there are three, are the main feature in this building, and hold an important, not to say obtrusive, position in the centre of the room. There is not much to be seen in them, however, big and imposing as they are. A circular glass window on the face of each shows inside a number of dials similar to, though much larger than, the ordinary gas-meter for house use. A clock wags a sedate brass pendulum in the centre, and every hour a record is taken of the quantity of gas manufactured.

I got the horrors looking in and watching the tens - of - thousands hand covering the ground. I tried to figure out a quarter's bill at the ordinary rates, based on a calculation from five minutes' observation. This depressed me and gave me a headache, so I quitted meters and mental arithmetic and went over and watched the test flame burning steadily away up in one corner of the room in a polished little case with a glass door to it.

Mr. Wright did make me understand the working of this contrivance; but headache left me hazy on the subject when I tried to remember it. It is a Governmental regulation, any-way, that a sixteen candle-power flame shall be seven inches in height, and in this arrange-ment the test can be applied and the quality of the gas determined.

Thus tested and measured, the gas once

more travels down the yard to one of the
numerous "holders," as they are termed—the
familiar great iron tanks supported in a circle
of massive columns, up and down which guide-
wheels travel and stay the " holder " from
getting jammed, or rocked by the wind-
pressure that in a decent breeze on such a vast
surface must be enormous.

They have quite recently set up a veritable
giant even amongst those giants in Gas-Land,
and I may as well describe that one, for they are
much of a muchness in constructive arrangement are these "holders," and differ only in
size and storage capacity.

The great fellow I was inspecting is made in
three sections, which fit telescopically one
within another. A third of the height, therefore, represents the depth of the tank-well in
the ground into which they subside.

This is an ordinary well, brick-lined, and its
chief importance consists in the fact that it costs
something like £10,000 to dig and construct. The
" holder " itself runs the company into another
trifle of £15,000 or so. Considered as a monu-
ment, it would doubtless be imposing and impressive, but something quieter and less costly

would satisfy me if I had that sum lying about
and getting in my way and irritating me so
that I had to put it out to service.

Water is an absolute seal for gas, and is
used to prevent the escape between the loosely-
fitting sections and from between the "holder"
and the tank. In the centre of the "holder,"
rising up from the ground to a few feet above
the water-level, are set two capacious pipes,
one for filling and one for drawing-off. The
gas is pumped in, and the pressure causes the
inside receiver to creep slowly upwards, the
guide-wheels holding it in place and allowing
it to travel freely. When it has fully expanded,
an outward-curved rim catches under a similar
rim curved inwards on the second tank, and
that is slowly dragged upwards, until the third
section is reached and hoisted in its turn.

The "holders" are constructed of riveted
iron plates, of half, quarter, and eighth-inch
metal, with a domed roof.

A fussy little steam-boiler was at work at
the time blowing steam into the water contained in the basin. I gathered from Mr.
Wright that the frost would be apt to com-
plicate matters, and this operation has to be

kept up all through the severe weather to prevent the water freezing.

The gas is compelled to travel back once more to the meter-house before they have quite done with it, where it passes through one of a series of big iron governors, of which there are five—one for each district—that by rising and falling over the outlet-pipe serve to regulate the pressure, in much the same manner that the governor of a steam-engine equalises its speed. And so it travels out of Gas-Land into the street, and I, having seen all there was to be seen, and having missed hearing nothing for the want of asking, wrapped my coat-collar up round my ears and travelled out with it.

IN PAPER-LAND.

PLEASE, I've come."

It was rather an idiotic sort of way of introducing myself; but then, I had obtained a kindly and prompt invitation from Mr. James McMurray, the managing director, and having previous acquaintance with Mr. Boyd, the manager of Paper-Land, I had even gone so far as to call upon him to take notice that on such-and-such a day, weather and tide notwithstanding, I should present myself at the McMurray Paper Mills, Wandsworth, at the hour of three in the afternoon, and that he was there and then to hold himself in readiness to give me service. I wanted to be "personally conducted," I told him, and he was to be the honoured victim. He stood the shock stoically, and even smiled and seemed pleased to see me when, at the appointed hour, I stuck my head in the doorway of his snug little sanctum just within the gates of Paper-Land, and said:

" Please, I've come."

We told each other how cold it was then, and how cold it was the night before, and how cold it would be the next day, while Mr. Boyd collected his Arctic outfit together and prepared to brave the inclemency of the weather outside.

First we took a walk into the country, and it was very charming country, too, for all that it wore such a wintry aspect. A broad, open, snow-covered stretch of land lay in front of us, with clumps of trees and bushes in the far distance; in the near foreground large filter-beds frozen over and suggesting a fine opportunity for a gay throng of skaters to disport themselves merrily could they get at it—but they can't; while right at our feet the silvery Wandle ran its course with a pleasant ripple and splash as we paced the broad staging that embanks it.

That walk was chilly, but bracing and vastly

pleasant. My industrial touring had, hitherto, led me into lands not picturesque, certainly,

A DUSTER.

and possessed of nothing in the way of scenery capable of appealing to the artistic sense that

8

I keep concealed somewhere or other about me for state occasions.   But Paper-Land certainly is prettily placed, and struck me as being a cleanly and pleasing locality altogether.

The object of my walk, I discovered, was this. They use a vast quantity of water in Paper-Land, and Mr. Boyd, with kindly forethought, had figured out that this fact might come upon me in some unguarded moment later on in my touring, and knowing from painful experience the likelihood of getting frozen out, I might conjure up visions of the awful results if such a calamity overtook Paper-Land, and go and faint in a tub of pulp and spoil the men's work. Having seen the water supply for myself, I should not be likely to entertain any fears of that sort, it was reasoned, and rightly.

The water is pumped from the river into the filter-beds in the same manner that our drinking supply is treated, and is drawn off from these for use in the mills.

We turned back and entered the engine-house, where I watched, with a species of fascination, a marvellous great monster exerting its eight hundred horse-power capacity for the utility and well-being of Paper-Land.   Big and

capable as it is, however, it has not the monopoly of the motive-power, which may save it
from any latent tendency to uppishness.

These mechanical monsters are wonderfully
deceptive. So silently and easily and unostentatiously do they do their business that they
inspire the contempt of ignorance. You feel
as though, if it were not for dirtying your
hands, you could take hold of the crank yourself and spin the huge wheel round, and the
engineers wouldn't know anything about it
until they came running in to see what was the
matter with the engine, that it travelled so
much faster and better than usual.

The boilers, which number eleven, of late
design, and fitted with mechanical stokers and
every other device that will save labour and
ensure good working, are bestowed in two
buildings, and in visiting them I discovered to
my satisfaction that I was making my last
expedition out of doors.

Both buildings presented a remarkably clean
and orderly appearance, and were so snug and
warm that I felt, as I walked round and asked
stupid questions indiscriminately of anyone
near enough and good-natured enough to

answer me, that I should rather like to be a stoker myself while that weather lasted, with the mercury endeavouring to shrink through the bottom of the thermometer.

Some of these boilers create steam of a superior standing, so to speak. For instance, it does not have to shove a wheel round, the common lot of steam created by boilers, which must be very monotonous and unsatisfactory work, but is blown direct into the mills to cook grass and heat cylinders, in a manner that I will presently describe.

. And so I came out of the boiler-houses and went straightway into the covered realm of Paper-Land.

Now, before I begin here, I must state, however much it may pain you to hear it, that paper is paper, whether it be paper to wrap up parcels; paper for irresponsible persons like myself to scribble upon and beautify for ever and ever—the compositor is laughing, I can feel him doing it. He is a mundane creature, and he cannot rise to the perception that a beautiful thought expressed with a J nib and glutinous ink is not the less beautiful that the page containing it contains also five blots and

THE BOILERS.

a smear made with a coat-sleeve roughly drawn
across wet ink, the impression of hot tobacco
ashes, and letters so curiously impartial that

they are ready to become whatsoever the reader wishes—or whether it be the broad printed sheet of *The Family Circle* that carries the same thought out into the world to beautify the homes of thousands of men, tens of thousands of ladies, a-wearied with the domestic afflictions of the day, and millions of innocent little children, laying up on the sofa, now that the day is over, recuperating after the fight with Billy Stiggins, or resting from the fatigue of snowballing crotchety, red-headed Sam Jones, the village snob.

After this long digression, I will repeat : paper is paper. The process is the same, broadly speaking; only the materials and the ultimate purpose differ. Paper-Land is chiefly concerned with the product for newspaper work, and incidentally I discovered vats full of *The Christian World*, *The Illustrated London News*, *The Times*, and sundry other weeklies and dailies. They were all in a very elementary condition, I must say. In fact, I didn't recognise them at all, and when Mr. Boyd pointed out a tub of *Christian World* to me I was almost overcome, it struck me as so pathetic.

"Why," I sobbed, endeavouring to extricate

my handkerchief from my pocket—a difficult
task when you are wearing padded gloves—and
dropping my note-book into a tub of *Literary
World* standing just behind me, in the attempt,
"I t—took it for a tap—tapioca pudding."

This occurred during my peregrinations; but
I have mentioned it at this stage *à propos*. It
was the only pathetic incident that occurred to
mar the interest and enjoyment of that memorable tour.

Up, up, up we went, and finally I stood in
an immense store-room where the raw material
is stacked up ready for immediate use. Piled
up in corners and round about the walls in iron
and rope-bound bales it stood—grass. Not the
sweet, soft, short, green grass of English
meadow-lands, but a long, spiky, tough material that hails from Africa's coral strand and
the sunny land of Spain. "Tripoli" and
"Esparto" are the technical designations; but
I detest technicalities and ignored the information.

It is a tough, tangled, and altogether unpromising-looking material, but with care and
kindness they manage to reduce it to perfect
submission and bend it to their will.

Like unruly children, the softening process is best commenced by a thorough beating, to knock the nonsense and the dust out of it, and everything is done for the best in Paper-Land. To this end it is wheeled into a large room, in which the ears are bewildered by an awful clash and clatter, and the eyes are smartened in one sense by the sight of a number of huge horizontal drums, from which endless bands of broad, stout canvas continually bear away the dusted material into an adjoining room, and in another by the clouds of dust that rise continually into the air. An operative will step up to the bale and sever the bands with a business-like looking pair of shears, and the grass is then fed into the "dusters" by women who pitchfork it about in the old familiar hay-making fashion.

A toothed drum, revolving rapidly in the dusting machines, tears the grass apart and liberates the dust and débris, which is sucked down and away by powerful fans operating below.

It is assisted from the "dusters" by a large wooden comb affair that drags it forward and pushes it on to the canvas platform that, con-

PULP-MAKING MACHINE.

tinually travelling, leads it up and **away** through an opening in the wall and precipitates it into the boiling-house.

I got into the boiling-house in a more prosaic
and commonplace way—through the door—
and beheld—well, not much. The place was
full of steam that smelt of stewed vegetable
matter and showed nothing through it, and
the general suggestion was that of a washing-
day on a large scale.

I have a suspicion that these machines are
not at all discriminating in their action, and I
didn't exactly yearn to become paper. Such a
fate would be doubly hard, for as " cream-
laid " I had an idea that I should prove a
failure, while for all practical and useful purposes as a scribe I should be spoilt by the
process ; so therefore I kept closely in the rear,
and allowed my guide the risky honour of
precedence. He took it without flinching, and
led me by vast domes of iron fitted with a
bewildering display of pipes and taps and tubes.

Emerging on the other side, I discovered that
we were standing on a floor built almost level
with the tops of immense vertical boilers, the
great domes of which were showing at intervals
through the flooring. The manhole of one was
open, and I leant over and peeped in. Far
below me a man was emptying the cooked

material out with a pitchfork, and as I watched him, Mr. Boyd explained the process to me. These huge boilers, each holding three tons of material, are filled with grass from above and a supply of water, along with a proportion of caustic soda. The manhole is then shut and screwed down, and a valve operated from above admits a current of steam, and the grass is left to boil.

When they think it is about done, they turn off the steam and cool down the boiler with a current of cold water, and empty the contents from below into tubs that run about on wheels —that is, when they are pushed they run about; otherwise they stand still on wheels. The boiling process has the effect of destroying the silica and softening the material, which by this time has acquired a brown, flabby, and somewhat dejected appearance. It is used to a warm climate, but it has begun to conjecture that the temperature of Wandsworth is too enervating, and yearns once again for the tropical chilliness of African tennis-lawns.

Not much time is allowed for vain yearnings, however, and even before they are formulated the grass is on its way into another building,

where it is washed and broken up—ground up
—beaten up—which you choose.

Beaten, boiled, washed, and broken up!
Truly, the discipline is Spartan in its severity.

Huge round tubs, into which taps are pouring a continuous supply of water, are ranged
in long rows up and down this room. These
contain sets of powerful knives, which work
after the nature of a grass-cutter, and machinery which keeps the contents continually
revolving. "Washers and breakers" they
term them. After passing through a "breaker"
the grass is grass no longer—it is pulp, a
brownish, soft, slippery substance that is rather
suggestive of a bran-mash.

Valves let the pulp down into a similar
arrangement, set at a lower level, in which it
is submitted to a similar treatment, with the
addition of a mixture of chlorine in the water
to remove the colour. After this bleaching
process the pulp is of a beautifully white
appearance, and is suggestive of some special
dessert dish rather than grass. They go on
washing and churning it a bit after this, and
then by the opening of a valve it is allowed to
flow down into immense lead-lined vats, where

THE PULP FLOWING ON TO MAKING MACHINE.

it is thoroughly mixed with water by mechanical agitation. From this vat it is pumped up to what is termed the " presse pate," where it is met by a stream of water which carries it along in the form of very much diluted milk over a shallow, ribbed trough, that retains the heavier particles, such as scale or heavy grit, through strainers, which free it from oddments in the shape of roots, until the stream, freed from all impurities, finally passes through an immense pipe on to an endless band woven of a particularly fine wire, that carries it along, extracting the moisture as it travels in the manner that I shall presently detail, and coils it over and deposits it on to a table in a form suggestive of wet blotting-paper. This is a duplicate process to the actual making, and I thought at first that I was witnessing the operation, until I saw the operatives roughly folding the sheets up in irregular batches and depositing them in metal trucks by their side.

In these trucks it is wheeled down a long passage to the " beaters," where it is subjected to another washing and pounding in order to remove the chlorine that would, were it permitted to remain, prove detrimental to the appearance of the finished product. A certain amount of colouring matter is then added, and it is permitted to flow into the storage chest— an immense lead-lined vat, similar to the one previously mentioned—which it reaches in the form of soapy or milky water.

Along devious and strange turnings, and through unexpected and out-of-the-way doors, I reached the same goal, and beheld a wonderful spectacle—a tub of *Christian World*, in which any one not too fastidious might easily go a-swimming. From the storage chest the milky or soapy fluid is pumped into a duplicate in miniature just within the adjoining room, set at a height above the paper-making machine. And such a machine! Extending the whole length of an extremely lengthy building, it is bewildering in its multiplicity of mechanical marvels. Immense square bands of red rubber gauge the width of the sheet of paper that is being made, and keep the pulp within bounds. A very fine brass network platform, on to which the pulp flows, moves continuously forward through a roller that imprints the water-mark, rollers that give it a squeeze here and a squeeze there, and rollers that coil it upwards when it has become a self-supporting sheet of paper, dry it and mangle it, and wind it up for later treatment.

Vacuum boxes underneath the bed of this machine—the largest and most marvellous I have ever had the good fortune to view—operated upon by powerful pumps adjoining, suck away the moisture from the stream of pulp as it travels up, and leaves the soft and solid substance behind in a thin, even sheet that is amenable to pressing and process.

It struck me very much in the nature of a miracle, and the simple word " astounded " about expresses my feelings while I stood and watched that machine at work. Think of it for yourselves! At one end you see gallons and gallons of a soapy-looking fluid being poured on to the brass net, that bears it ever onward and still comes up, like Oliver, crying for more. As you walk down with it you notice that it is not looking quite so sloppy, and by-and-by it occurs to you that it has taken on the appearance of a rice-pudding somewhat badly constructed, and so you come to a pair of rollers, huge brass concerns faced with felt.

REELING THE FINISHED PAPER.

Presto! the pudding has become paper to all appearances, though it is not yet quite suitable

9

for practical purposes. And so it travels on until it is firm enough to leave the bed, and to commence a devious journey through a series of eighteen metal cylinders of gradually increasing temperature, while the steam rises in a cloud above them from the drying sheet, and a little automatic measuring instrument ticks off the mileage of paper—they make it by the mile, four or five being the usual length—and knives on each side shear the rough edges and reduce the sheet to the required width.

It takes a special engine to operate the making machine, and the clash and rattle prevented me from going into details that I should have liked to dwell upon; but I gathered that the paper is sized somewhere in the course of its journey, and what is termed " surface " is imparted to it by a set of chilled iron rollers in its passage from table to reel. The width made, I must state, is in most instances just twice that of the required dimensions, and as it emerges from the rollers a knife divides it down the centre and allows it to be wound up on to two reels that are fitted into a frame at the end of the machine. The paper is carried overhead to these reels, and you pass under arches of it as you leave the room on the way into the hot presses.

The process it has undergone is sufficient for some purposes, but where a high "finish" is desired, the reels of paper are brought into an adjoining room and put.through another series of steam-heated cylinders and cotton rollers, from which it is reeled off on to spools or cut into sheets, and is ready to pass out of Paper-Land, to testify to the merits and swell the profits of its birthplace.

At the extreme end of the finishing-room— which, by the way, is on the ground floor, I, to my surprise, having travelled downward through the whole length and breadth of Paper-Land—stands a marvellously perfected piece of mechanism that unwinds reels and reels of paper, and cuts them into sheets of regulated size. It is not every journal that can aspire, like *The Christian World*, to the dignity of printing its edition from four-mile reels—that betokens a good standing, large circulation, convenient size, elaborate machinery, and a host of other signs of a sound and flourishing condition. Smaller fry find it more practicable to print off the sheet, and this ingenious contrivance was meeting their requirements with an adjustment and action so perfect and so true, that only two "hands" —representing four arms, be it remembered— are required, and they simply square up the sheets as they are sent down along an endless revolving band of broad felt.

The daylight was fading rapidly, and the electric light, which is fitted throughout Paper-Land in office and in mill, was enlivening the scene as I made my way past and round reels of paper waiting to be forwarded to their destination up in London printing offices.

And so I gathered up my umbrella and, removing the débris and dust that I had acquired up in the "dusting" room from my wardrobe, I thanked Mr. Boyd for his courteous and kindly efforts on my behalf, and wended my way down to the station.

I had a curious dream that night. I dreamt that I was hungry. I went into Gatti's in the Strand to get something to eat, and they brought me, of all things in the world, a sago-pudding. It wasn't an inviting-looking pudding; it seemed badly constructed to my mind. But they assured me that it had been

thoroughly dusted and that they had extracted
nearly all the chlorine from it, so I sat down
to eat. Before I could touch it, however, it
was whisked under the table and came curling
up the other side in a broad sheet of paper
on which was scrawled in "spooky" hand-
writing: "Haven't you nearly done?" And
I answered and said: "I have!" And I have.

IN SOAP-LAND.

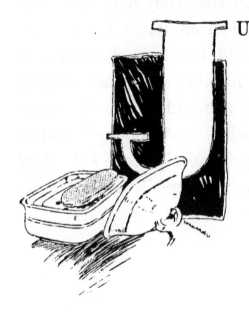

UST past Bow Bridge, on the left-hand side coming from the station, the observant traveller will notice the narrow and somewhat insignificant turning down which I wended my way in quest of Soap-Land. Cook's Road it is termed, and it has its termination in a pair of huge gates that shut out the public at large, and give admittance to the colony, covering fifteen acres of ground, and containing over three hundred inhabitants during working hours, that I had come to explore.

I was " on business," and no difficulty was therefore experienced over entering a land that is jealously guarded by day and by night, and my credentials carried me through to the general offices, where I waited the advent of my guide that was to be.

This gentleman presently put in an appearance, and, introducing himself as Mr. Alec Cook, apologised pleasantly for his business-like "get-up." He was wearing the white smock affair that the inhabitants of these lands adopt to protect their clothing from tallow splashes and grease marks generally, and he explained to me that he was practically superintending the finishing of a "copper" of soap, such being the universal custom in the firm of Messrs. Edward Cook and Co., the owners of Soap-Land.

Through the offices we walked together right into the factory, where my guide left me for a moment while he went back to give some directions concerning the work he had abandoned on my account, and I gazed out on the River Lea that, running below, serves Soap-Land nobly in the matter of bringing to and bearing from easily and expeditiously the raw materials and finished product. The Great Eastern Railway trains are thundering past on

the right-hand side, and I realise something of
the advantages that have induced Messrs. Cook
and Co. to cover the greater part of fifteen
acres of land in a densely-populated and unat-
tractive suburb with plant and property for the
manufacture of washing material.

I gazed out on the frozen river and bethought
me of the poet's lines, "I saw three maids
come skipping o'er the lea," as I watched three
bargees reach themselves over the side of their
unwieldy craft and lounge heavily shorewards.
I should have gone on conjuring up poetic
fragments, only just then Mr. Cook returned
and announced his readiness to attend to me.

"That is our highway," he remarked, following my gaze—"that and the railway that runs
into our own siding, though we find sufficient
road-work to justify us in keeping fifty horses
and vans in constant use." Steam cranes hoist
the material on to the wharf, and a system of
rails serves to bear it to any part of the works.
The tallow and fats used in soap-making, however, find their way by a steam lift to the top
of the main factory, where they are started on
their evolutionary journey.

Another building across the way supplies a

A COPPER IN SOAP-BOILING ROOM.

great quantity of material direct, the fat being boiled down there by steam, and blown through a service of huge pipes into the making department. I discovered that process for myself later on, and contrary to expectation, it presented no particularly objectionable features.

Soap-Land is a land of complicated marvels, and never have I beheld such a bewildering entanglement of pipes and valves and tubs and vats as I ventured myself amongst for the next two hours. Huge metal tanks for storing fats greeted my eye on the upward journey. During the winter the softer fats can be handled in barrels safely enough, I gathered; but in the warmer weather they are better confined in these drums, where they don't get so much chance to ooze through and mess up the floor. On this floor the barrels of tallow are adjusted over a copper pipe projecting into the room, and a current of steam is blown into them. "Phew!" remarks the tallow, "this is getting warm. I'm going to run!" And it does, through openings prepared for it down and across to the soap-house on the floor below.

Immediately under this part of the reception-room, as I may call it, are the alkali pans. My guide called them "pans," though such a simple and unsuggestive term hardly does justice, to my way of thinking, to the immense tanks in which the soda is dissolved.

Fat by itself is not a possible washing material, whatever may be stated to the contrary; and to make soap alkali of some sort must be incorporated. Carbonate of soda causticised by lime is the most general alkali used in Soap-Land, and the care and proportion used in mixing constitute the difference between a good soap and a bad one.

Leaving the giant pans of soda, in case the men might want them, I turned away through a door on the right and entered the soap-house, a tremendous building with an open space above running right up to the skylights. Ranged along the wall, all down one side of the room, are the soap-coppers, of which there are twelve, and they were a revelation even to me who seem to have been spending my declining years in inspecting huge storage contrivances.

They are all immense, but the smallest, which only holds about ten tons of soap, could be hidden away in the really large fellows, and the men would spend hours looking for it and never dream of its presence so near their feet and yet so far. When I instance the fact that one hundred tons represents their capacity, you will be able to figure out for yourselves some idea as to the size of these monsters. Of course, you do not realise this fully even from an actual inspection, unless, like myself, you are fortunate enough to catch one empty.

They are built down through the floor, and you are only able to see the three or four feet projecting into the soap-house, over which the workmen are easily able to manipulate the contents; and filled with soap, of course there is nothing to give you a sense of the real depth of the bubbling, yellowish, creamy substance. Clarence was drowned in a butt of wine, and it seemed to me that it would be just as easy for a man—duke or commoner—to get drowned in a "copper" of soap, only I was satisfied to leave the record that that unfortunate noble-man was at such pains to establish unbroken, and refrained from making the test.

The third or fourth copper, I forget which, along the line was empty, and peering down into its depths, it struck me more in the nature of the hold of a ship than a mere soap-constructing tank.

THE ALKALI PANS.

In the making of soap the fats and tallows are run down into these coppers in the manner

I have before explained, a due proportion of
resin is emptied in with it according to the
nature of the soap being made, and the whole
is boiled up for several days by a current of
steam blown through it. This process, by the
bye, is termed wet-steaming, in contrast with
the more familiar method of dry-steaming,
where the steam circulates through a system
of pipes without finding its way into the material
it is melting.

When they judge the fat to be of the right
consistency, a valve governing one of the pipes,
leading from the alkali tanks above, is opened,
and the stream of soda solution immediately com-
menees its work of transforming the melted fat
into soap by a sort of magical process. The
surface is first converted, and then the lower
strata take their turn in due rotation.

Salt liberates the soap that, probably not
being so sturdily independent, is still held in
solution after the spontaneous separation has
taken place, and the partially spent "lye," as
the soda solution is termed, is used again and
again for cheaper varieties of soap until it is
fully "spent."

And so the soap is made by a process as

simple as it is marvellous, although, of course,
consideration has to be given to the preliminary
operations that require a vast amount of
knowledge, skill, and forethought to bring
about anything like a satisfactory result.

"I suppose," I remarked, tentatively, to Mr.
Cook, "that it would be easy to spoil a 'copper'
of soap?"

"Remarkably easy," he replied, drily. After
that I didn't like to ask him to let me make a
bar, even of some common kind, all by myself,
to take home with me as a proof of my clever-
ness and skill, around which, as the years
rolled by, I could pile up history and romance
according to the believing capacity of my
audience.

I did not see the actual soap-making process;
but I could follow it clearly enough, thanks to
my guide's lucid and business-like explanations,
and as I came away from those coppers full of
tons of "primrose," and "yellow," and "curd,"
it occurred to me that when literature has
become a lost art, and my pen is idle for the
want of something to do, I could put it aside
in favour of a white blouse and ladle, and go
and be a jolly soap-maker.

Before leaving the soap-house, I must dwell once again on the wonderful system by which the coppers and pans are connected together. Complicated as it necessarily is, it is so complete that the melted material can be transferred to any tub or vat in any required part of the building by simply turning a tap, or by putting a pump to work. Pipes are everywhere, under foot and overhead, and it struck me that, by a little judicious monkeying, a simple sort of individual like myself—not necessarily wicked, mind you, only curious—could do more damage in ten minutes than could be righted in a week. As soon as the soap is formed, it is sent travelling along troughs into the frame room by the powerful action of a large double-cylinder pump.

The idea of pumping soap about struck me as funny, and I laughed my hat off, and went down into the frame room after it with more celerity than I had intended.

A frame is simply a huge iron mould capable of holding twelve hundredweight. The four sides fit together like a toy money-box, only, of course, there is no lid, and are fastened in their place by iron bolts. When ready for

IN THE FRAME ROOM.

filling it presents the appearance of a deep,
oblong tank of roomy dimensions; and I had

no idea at first that it was constructed in sections, although precisely how, having got the soap in, they were to otherwise get it out again didn't strike me. The frames are set in regular rows up and down the room—there must have been hundreds of them — and running just overhead the troughs that I have alluded to were suspended. A hole in the bottom of these troughs over each, which can be plugged when necessary, gives the right of way into the frames, so that when the pump is set going the stream of soap runs the length and breadth of the room, and the frame filling goes merrily on—a simple yet ingenious arrangement that saves untold labour and time.

The soap is left in the frames to set and harden, when the men come round and remove the casing, and expose a hard, solid, twelve hundredweight block all ready to be cut up into slabs. This cutting is done by a strong man and a thin steel wire, and they do it remarkably well between them. Hard work it showed itself to be, too, and when my guide invited me to test the resisting capacity of one solid-looking slab, I declined promptly. I have cut up cheese by the same simple means; but cheese isn't soap, and the work looked too solid for a mere after-dinner diversion. Now, according to the nature of the soap being dealt with, the subsequent treatment varies; but the beginning is in those slabs that the man and the steel wire between them are perpetually preparing. Through a devious pathway, strewn with pipes and littered with pumps and tanks and machinery, I found my way—or rather Mr. Cook found it for me, for I would engage to get promptly and hopelessly lost inside of two minutes if left to myself—to the stamping-room, where the tablets and cakes are prepared, after being cut into bars from the slabs down below.

Both hand and machine moulds are at work stamping and pressing the tablets into shape; but the hand-machines are much preferred, I gathered. A metal " die " gives the impression in very much the same manner that the coins are made at the Mint, the soap being previously oiled to prevent it sticking.

After that I should be afraid to attempt to tell in detail what I did see. I saw so much that my brain got confused. The contrivance that attends to my thinking department grew maudlin, and left me an easy victim to the mesmerising influence of tablets of " Riviera," "Savon de Luxe," and bars of " Lightning Cleanser," and other domestic and useful varieties.

As I turned away from that busy scene of stamping, wrapping, and packing, I bethought me of " ye legende of Bowe in ye east," a pretty story, and I shall just have time to tell it you while the lift is bearing me down to other strange and unfamiliar scenes. Only, mind, I don't vouch for it.

Years and years ago a colony of bees swarmed across the marshes and settled down near Soap-Land. They were steady, sober, hard-working bees, and their honey was the best that could be obtained, so that people flocked from the uttermost parts of the suburbs to get it, and the bees put another halfpenny a pound on to the price, and prospered exceedingly. They began to grow conceited over their industry, and at last got so uppish that they came to look down on the bees of other hives. They were thick-headed sort of bees, and they plodded away at their toil, and never supposed that any of their tribe before or since had or could become such champion honey factories as themselves.

One bright summer's morning, however, one

CUTTING UP THE SOAP.

of their number flew by chance into Soap-Land, and came unto the stamping and packing room. It felt sick and dizzy as it watched the scene.

Verily, what were their puny little efforts compared with these? It crawled feebly downstairs, and beheld oceans and rivers of—what? It was yellow soap; but the bee was a thick-headed little creature, and knew it for honey. It didn't do any more work that day, and when it reached the hive late that night it had a headache. The next day it took a friend over to see its discovery, and the news then very quickly spread all over the town, and the bees from other hives came over and flew about in groups, and said nasty, spiteful, and irritating things. An investigating committee from the hive went over to Soap-Land to see for themselves whether they were being "had," and when they saw how really insignificant they were they lynched the discoverer on the spot, and then went back and closed down their works and became dissipated and abandoned. Like Othello, they deemed their occupation gone or superfluous, for it didn't seem worth while to be making honey by the half-ounce with industry and moral integrity when it was being made across the way with machinery by the ton. The moral seems to be, "Never introduce your donah to a pal."

I got to the moral and the ground floor at about the same time, and found myself in a veritable room of soap.

There were bricks and mortar and wood-work of course; but stored up in immense bins around the walls were tons and tons of soap. It was stacked away in bars, and the cutting of these bars from the slabs I had seen turned out in the frame-room was as simple as it was in-genious. It was done in a wooden frame affair that lifted the slab and forced it against a row of steel wires set at regular intervals along the back.

Bin after bin of soap was opened for my in-spection, and I walked round and tried to form an estimate, so that I might give you all some idea of the quantity by a comparison that would do justice to the subject without straining your faith in my veracity.

Simple statements are safest, I reflected, so I will confine myself to informing you that one of these bins will stock fifteen tons of soap, and the bins are numbered by the dozen.

Up to the first floor again by means of the lift, and we emerged on to a light but substan-tial ironwork bridge that led into another soap-house, where special soaps of various kinds, including mottled and soft, are prepared by processes not greatly differing from those I had witnessed in the other building.

Recrossing the bridge, we sought the ground-floor again and came out into the open, where I had an opportunity of inspecting the main-springs of the place, so to speak. The boiler-house was very much as other boiler-houses, containing the six immense Lancashire boilers that supply steam for the melting tanks and coppers, as well as for the engines, of which there are many. A small percentage of it is used to work the lifts, the business end of which I passed on my way over to the stables. There was nothing much to see; but the method of working is ingenious. An immense iron cylinder, weighted with seventeen tons of metal, and bearing on a piston-rod, is upheld by the pressure of water forced under it by a steam-pump. When the lift is wanted on the upper floors, the men open a valve which admits the compressed water, and allows the cylinder to descend, and, as that falls, so the lift rises until it reaches the limits of its possibilities, when the pump sets to work again to reinstate

STAMPING OUT TABLETS.

the cylinder in its original position ready for future action. When the men want to descend, they merely let the water out of the pipes, and the lift falls by its own weight, and slowly and orderly is the fall thereof.

The stables present several novel features, and are pleasing examples of the care that a big firm not only does not disdain but considers necessary to bestow on its dumb *employés*. Space is economised by building in two stories, and the dwellers on the upper floor walk up an inclined path to their beds with all the alacrity of a good Christian who has done his duty through the day, and feels that rest and repose have been fairly and equitably earned. Fitters' shops, where all the repairs to tools and accessories, boilers, coppers, and pipes are attended to, saw-mills and box-making departments, and a farrier's shop were all passed and inspected by me on my way round to the most interesting department of all—the perfumed soap room.

A twenty-pound note doesn't go very far in some of the scents used in these soaps. In fact, a four-ounce bottle would provide ample accommodation, which seems extravagant to a man with a bias in favour of obtaining quantity for his money. You will readily understand, therefore, that it would hardly do to subject such costly liquor to the evaporating and wasteful effects of heat. The cold process termed "milling" is therefore adopted, and for its operation a special plant is required, and, of course, only the high-class toilet and antiseptic soaps are so treated.

Into the "milling-room" in which I was then standing, the soap is conveyed from the main factory in firm, white, and dense bars. The first thing to be done is to shred it in a machine that cuts it up by a coarse suet-chopping process. These shreds are dried and are then transferred to the "milling" machine along with the perfume that it is desired to add. Immense rollers grind the mass thoroughly up together, and let it out underneath in the form of long, thin, perfumed shavings. Antiseptic ingredients are incorporated in the same way. In order to get these shavings to bind together in one solid bar again great pressure is required, and this is applied by means of an Archimedian screw working in a sausage-machine sort of arrangement. The shavings are fed into its capacious interior, and are forced downwards to a narrow outlet. At first a round mouthpiece is used, and the soap issues

in a long, thin tube, which is broken off and
thrown back into the machine until it is judged
that the proper firmness and density has been
reached.   Then the mouthpiece is removed,
and the soap issues in one continuous oblong
bar, which is cut up at regular intervals by the
operative in charge, and taken away to be
" tabletted " and stamped.

An ingenious and a pretty process has been
recently adopted in Soap-Land for indelibly
lettering these tablets, and the idea has "caught
on " extensively.   Hotels were well in evidence
among the specimen tablets I saw.   I expect
they find it a preservative as well as an adver-
tisement, for it looks bad to be found using a
cake of soap marked " Timson's Hotel," when
Timson's Hotel is many miles away, and the
user does not answer to the name of Timson.

The lettering is done by means of needles
set in a stamp.   It is a messy operation, but
very fascinating to watch, and is chiefly in the
hands of young girls.   A supply of some harmless crimson or blue colouring matter is automatically emptied from a trough over the
tablet, and runs into the holes made by the
needles.   The superfluous colouring is then

MILLING MACHINE.

washed off by hand, and as long as the tablet
lasts the lettering can be read.

In this room I was shown trays and boxes
of the different soaps made in Soap-Land.
There are something like one hundred and fifty
varieties. Some of the tablets were real works
of art, and it seemed almost a sin to use them
for common washing purposes. I have one
rose-coloured oval specimen by me now, with
singing-birds and rosebuds and other children
of nature stamped in relief on its surface, and
I cannot quite decide whether to put it on a
plate in the window and raise the artistic tone
of the neighbourhood, which is somewhat flat,
perhaps because it is beyond the ken of the
County Council, or to hang it in my study as
evidence of æsthetic culture.

"That is germ soap," Mr. Cook remarked,
noticing that my attention was centred upon a
box of pretty little square tablets.

"I didn't know germs used soap," I replied,
with a look of innocent wonder.

It was a slip on my part. I am, as a rule,
very careful about giving myself away. People
generally mistake me for Irving, except the
younger and more sentimental, who incline to
the belief that I have either got the earache or
am about to be married.

My guide only laughed, but I could see by his manner then that he had fathomed my secret and knew me for a humorist, a heartless creature who could even laugh at germs.

I gathered from his explanation that they—the operatives, not the germs—put bin-iodide in these tablets of antiseptic soap, and that the germs don't like it. They die when they see it coming and turn to butterflies, and fly away and live happy ever after.

It was getting dusk as I left the perfumed soap room, and, unwilling to trespass wantonly and unreasonably on my guide's time and attention, I took a passing glance at the laboratory, stocked with delicate and mysterious appliances for the testing of raw material and finished product, and at a smaller chamber adjoining, where hundreds of pounds are stored up in tanks and bottles in the shape of perfumes, and came away from Soap-Land with an increased respect for those whose duty and pleasure it is to help to maintain the world's inhabitants in the state that is next to godliness.

IN POTTERY-LAND.

WHEN I set out to explore Pottery-Land I wondered whether I should have any difficulty in locating it. Lambeth is a quarter that I am not exactly familiar with, and I was afraid at first that I might overlook it, not knowing it, and, after wandering round and round all day like a superannuated babe in the wood, be discovered late at night, wrapped in despair and an overcoat, on one of the seats on the Albert Embankment. My fears were groundless, however; and, later, when I had whiled away half-an-hour or so in making myself familiar with the neighbourhood, I came to recognise that the real difficulty is in missing it.

Of course, the good folks in Lambeth don't know it as "Pottery-Land." They call it "Doulton's place," and the big gates and the palatial buildings that enclose and comprise it are variously inscribed "DOULTON AND Co., LAMBETH."

Mr. Rix, the art director, had kindly undertaken to guide me, and to him, at the appointed hour, I made my way, and, in passing by spacious offices and busy workrooms, I came to form some little idea of the magnitude of the task I had undertaken.

What I didn't realise in this respect Mr. Rix quickly made me sensible of. From him I gathered that Messrs. Doulton and Co. possess about half-a-dozen "Pottery-Lands" scattered about in Staffordshire, Warwickshire, and Lancashire, one at Paisley, and one in Paris, in addition to the one that I was about to explore; and that everything that can be manufactured out of clay, from a tea-cup to a two-thousand-guinea art vase, including bricks, blacking-bottles, and drain-pipes, is manufactured.

I reflected then, that if I set out to do this thing thoroughly, I should be discovered years and years hence by a search party, sent out from London to find me, a decrepid old wreck compounded of clay and weariness, taking down notes in the pantile department.

Something more rapid was what I really wanted, so Mr. Rix kindly consented to omit all duplication of processes, and on that understanding we left his office and went out into the street. Turning off to the right on reaching the Embankment I first discovered how Pottery-Land received its supplies. Dug from the pits in far-off Dorsetshire and Devonshire, the clay is brought up the river and transshipped into barges in roughly - squared portable blocks, which do not call for any delicate and careful handling that would, perhaps, be difficult for the average bargeman to readily acquire.

Pottery-Land possesses its own docks, and the entrance thereto runs through tunnels leading from the river under the Embankment.

Elderly maiden ladies have been known to express astonishment and some alarm at seeing a heavily-laden barge run into the massive

stonework and disappear from sight. They
have even conjured up awful visions of barge-
wreck, and have gone and disturbed some
passively - ruminating policeman with wild
requests for lifeboats and rocket apparatus.

Unloaded from the barges, the clay is stacked
up against the time when its services will be
needed. It is stacked up against a wall, too,
for additional support. It seemed to me that
this delay must have a deteriorating effect on
the material, but my guide assured me that
this was of no consequence, because they preferred to add whatever moisture was required
themselves, which relieved me immensely. I
take a great interest in these processes, and
when I think I have discovered an oversight
anywhere I usually point it out. The information, I reflect, may be of extreme service; but
as a rule it isn't.

The clay blocks are first of all ground into
powder, along with a proportion of old broken
stoneware, in a capacious iron tray in which
two heavy wheels are revolving. A fine dust
results from this crushing process, and this is
drawn into another machine, where it is mixed
with water fed into it with care and judgment

and a tap, by an operative whose discretion
can be relied upon to compromise somewhere
between a badly-digested mess and a mud-
puddle. When it is judged to be of the right
consistency it is scooped up by a number of
buckets, fastened to an endless band and acting
on the ordinary dredge principle, and thrown
into a gigantic sausage-machine sort of arrangement, where it is cut into pieces by revolving
knives. Issuing again in solid bars, it is
packed on to trolleys and run up to the floors
above by lifts, and stored in bins and cupboards
until required for use.

Coming away from the mixing and preparing
mill, my guide pointed out to me quite an
insignificant-looking corner where Pottery-
Land had its origin, starting modestly with a
couple of kilns and a dozen "hands." Over
two thousand "hands" are employed there at
the present day, and as many again in the
dependencies, so to call them, which goes to
prove that ill-weeds have not the monopoly of
growing apace.

Having seen the clay prepared and brought
to a portable, plastic, and workable condition,
I mounted to the floors above by leisurely

PREPARING THE CLAY.

stages. An Alpine enthusiast could here get some good and congenial exercise; but the mere seeker after truth and trade mysteries only longs for a balloon, or a stock that he doesn't happen to be holding, or anything else that has a tendency to go up without the exercise of leg-power. Once, a long while ago, I used to have something unusual in the way of soaring ambitions, but it never got what I call a fair opportunity to soar; the opposition was too marked, and it wilted and pined away from vexation and disappointment. On the fifteenth floor—I counted fifteen, but I may have been misled; these places are very confusing—I suddenly bethought me of it, and wished that I had brought the poor thing with me and given it the chance of one really good flutter. I would have waited for it down below in the basement.

Of the extensive variety of articles made from clay in the section of Pottery-Land that I found myself in, some are moulded and some are "thrown." If, for instance, they are making a kitchen sink they mould it. If it is a teacup they are wanting to make they "throw" it. It is a singular thing, but servant-girls unmake crockery by precisely the same means. They generally use the scullery floor, and rush in with the pieces and say, "Please, mum, the handle's come off." It is always the handles that come off. If jugs and teacups could only be made without handles people would get so sick and tired of seeing the same old bits of china about them year after year that they would take to smashing them with the kitchen poker in sheer desperation. The potter uses a wheel—I expect because he hasn't been educated up to appreciate the beauties of destruction, and can only grasp the minor interest of the mere constructive process.

On my way up I passed a room where the moulds were being made in plaster-of-Paris, and a little higher still I had an opportunity of witnessing an article—it was a kitchen sink— being moulded.

It is a simple process, and hardly needs detailing. The mould is made in sections, of an obverse pattern to the sink, of course, and is fastened together and kept in position by wooden wedges forced between the bands of thin iron encircling it. The clay, having been forced into circumstances that compel it to take the shape and semblance of a kitchen sink, is left in the mould to harden until it can be safely handled. It is then removed, and after being dipped in a bath of white glaze— that looks not unlike pipe-clay, and is in reality a glassy material that, fusing with the heat it is to be subjected to, coats the article with a hard, smooth, shiny, and impermeable surface —it is ready for the kilns; only, as I am not yet, it must remain in that stage to wait my pleasure.

It is a bewildering place is Pottery-Land! Trolleys bearing loads of clay are rushing along metal lines which seemingly lead everywhere about the building. Through the centre of several of the floors immense circular brick constructions, set about and bound with heavy iron bands and huge chains, rise upwards. You notice, if you are of an observant nature, that the floors terminate within a few inches of these constructions, and you peep through and see the ground-floor a dizzy depth below you. A considerable warmth, too, radiates from the massive walls, and this prepares you for the announcement that they are the actual kilns in which the pottery is " fired."

THE POTTER'S WHEEL.

Pottery-Land is practically fire-proof, and is specially constructed with concrete and iron-doors in case the twenty-one inches of brick-work in the kilns should prove insufficient to confine the devouring element within reasonable bounds. Nothing is attached to the kilns for another reason. The intense heat has an expansive effect on them, and they rise and fall with the regularity of a thermometer. The movement is infinitesimal in each section of the kiln, but many littles make a lot, and on the upper floors the displacement amounts to an inch or more. To hitch a floor on to that would be akin to the folly of building a dwelling on sandy foundations, which even Mr. Jerry, a celebrated and extensive builder in my neighbourhood, would not be guilty of. As a seesaw the construction would answer capitally; but for habitable purposes there would be drawbacks.

Stowed away in one corner, as though they were rather ashamed of it, I came upon a very interesting process by which drain-pipes are made by machinery. They use clay as well, of course, and a man or two just to keep things going like; but a sort of steam-hammer arrangement really does all the actual work, and, if I may say so without seeming patronising, does it remarkably well. Into the top of the machine men are continually at work shovelling clay. This is forced down by a ram through an orifice in the shape of a ring. A long steel bar rising through the floor is fashioned in the pattern of an ordinary drain-pipe socket, and after receiving this impression it drops again, and the pipe is pushed out of the mould with a curious sucking noise, and carried away to be finished off. They make something like fifteen miles of piping a week, bent and straight, and of all diameters. The finishing off is done by hand, the drain-pipes being placed on wooden frames that gauge the length, and the superfluous clay simply and deftly cut off with a thin copper wire.

And then I came to one of the oldest and the most fascinating of contrivances, the potter's wheel. All circular articles, from a ginger-beer bottle to a costly vase, are made on these wheels, and the process is termed "throwing," as I have before mentioned. It looked so delightfully easy and simple, and yet it is so confoundedly hard and tricky that, I gathered from my guide, a month's hard labour at it would advance the novice sufficiently to construct something that his fond relatives, with every wish to encourage him and do him justice, would recognise for a penny ink bottle if he told them that it was to hold ink.

The "wheel" is simply a metal disc which is operated by a belting from the shaft running down the length of the room. The ancient Egyptians lacked this convenience, I believe, and had to do their own propelling on the scissor-grinding principle, which couldn't have been so comfortable. The potters in Pottery-Land only use their feet to control the wheel by means of a foot-brake. After I had seen them working, I came to the conclusion that even with the advantages civilisation has given them there is plenty to tax all their skill and dexterity.

By the side of each wheel boys are busily engaged in kneading the clay with their hands to render it thoroughly plastic and free from air-bubbles, and as the potters are also needing the clay, they need to be pretty continuously engaged to keep the supply a little in advance of the demand.

ON THE LATHE.

The potter sits on his bench a little above his wheel, and taking a lump of clay he throws it down—hence the term "throwing"—on to the rapidly revolving disc. It is fascinating and funny to watch its subsequent evolution. Sticking a moistened finger into the centre, the clay begins to coil up round it, and spread itself out, and rear itself up again and "do tricks," all in obedience to the hand that is directing its movements. A couple of little forks terminating in pieces of whalebone stand out and give the only aid in the shape of a gauge that I could discover, and from these the potter knows when to commence bending over to form the neck. (They were making bottles when I witnessed the operation.) A flat piece of metal cut to the shape of the mouth is introduced, and then a few deft touches forms the lip, and the article is ready. Of the common variety of bottles a potter will turn out many hundreds in a day's work.

I was fascinated! The next time a person tells me that I "potter about"—people do tell me so occasionally, if they are related to me, or are sufficiently intimate to dispense with courtesies and speak their minds frankly—I shall say: "Have you ever seen a potter about turning a piece of clay into a bottle or a tea-cup?" Perchance the person will say "No." "Then," shall I answer him, "call me a 'messer' if you must be rude; but speak not slightingly of that which you know not of. It is sacrilege and silly as well."

Handles and spouts and etceteras are, of course, made separately. They are moulded or shaped by hand with the assistance of a knife and a bone stylus, and stuck on, the uniting surfaces being scored and damped slightly to make them adhere.

The potter's wheel was working in many rooms—more than I could possibly hope to visit, let alone describe—but description of a detailed nature is hardly called for. I saw them making teapots in one room and blacking-bottles in another, while a third was devoted to bottles for the reception of beer termed ginger, and a fourth witnessed the manufacture of big stoneware wine and spirit jars. The process in every case was the same, while the results were vastly and entirely different.

. I had some grand and beautiful thoughts while I stood and watched them making blacking-bottles, and I leant up against a bank of wet clay and wrestled with them, while the world went past me unobserved. There seemed to be fresh possibilities for the poet's Imperial Cæsar, dead and turned to clay, stopping a hole to keep the draught away, if he should by chance get brought in on a barge to Pottery-Land. It is true he might be fashioned into a blacking or a beer bottle—in the latter case, I suppose, he would become an Imperial pot—which would be funny, because he was undoubtedly a "pot" and Imperial when he was alive—but, then, he has as much chance of becoming a teapot, or even a vase of great value and choice workmanship. He might, if there is enough of him and he proves good clay, even fetch fifty guineas, which would be a good "first edition" living price for anything Imperial that ever walked over two legs and a subject's rights.

But how good and truly generous must Messrs. Doulton and Co. be! The average man, with a piece of clay and the choice between a blacking-bottle, which sells for less than a penny, and, say, a teapot, which will go from anything between half-a-crown and fifteen shillings, would elect to do the teapot. Not so

IN THE ART STUDIOS.

12

in Pottery-Land. They think of the needs of humanity's boots there as well as of its afternoon refreshment. But it is very commendable of them, and I should like it to be appreciated. These little everyday sacrifices for the benefit of mankind are not half sufficiently recognised to my mind.

The power of the potter is autocratic, and his intentions are only known to himself and to those in authority over him. When he picks up a piece of clay, does it know what it will become? Its parents and relatives see it set out with a lofty purpose and a burning ambition, and they talk significantly and mysteriously about it to their friends, who are thenceforth under the impression that it has won fame, and is to set the world agog over its "chaste design" and "delicate grace and beauty." Later on, they all meet in the kiln and recognise it for a blacking-bottle, and they snigger and make remarks, as friends will. We are very like that piece of clay, though, and Fortune, the potter at the wheel of life, is more curiously impartial than the potter in Pottery-Land. Of some of us she makes blacking-bottles, and of others Sèvres and Worcester, and then, before we are able to get baked, she will mould us up afresh, and the Sèvres will become a blacking-bottle and the blacking-bottle will become Worcester ware; and sometimes she flings us down and smashes us altogether, and will construct what the world terms "lamentable failures" out of the remnants.

I was going on in this strain, when my guide suggested that if I were nearly rested we might proceed, and thus brought back to a recollection of mundane things, I collected myself together, and was shortly after inspecting some very pretty art ware, in the shape of vases and decorative oddments.

After being "thrown" on the wheel these things are brought into the room in which I found myself, and are "turned" in a lathe just the same as a piece of wood would be. Mouldings and bands are cut on them during this process, and a desirable and even thickness is also obtained.

The articles are then taken away to the studios where, in a series of comfortable, quiet, and well-appointed rooms, a number of young ladies are busily engaged in artistically designing patterns on the soft surface. They are given an entirely free hand, I gathered, and the results struck me as being remarkably clever and tasteful.

The last stage of all is the kiln in which the pottery is "fired." I have already given you a rough idea of these kilns, which are simply immense bricked chambers, held together and strengthened by huge iron bands and chains. The fires play up the sides, over the top, and down through the centre again. At least, the heat and the smoke go down through the centre and away to the lofty shafts that tower over Pottery-Land like veritable giants.

A fairly large oval opening gives admittance to the kiln, and I was fortunate enough to see one being loaded up. The capacious chamber was about half-full, when I saw it, of any and every variety of stoneware. The articles are stacked in tiers, each layer being supported on movable slabs, which divide the kiln into a number of cells.

I bethought me somehow of Gray's touching lines as I beheld the men busily stowing away sinks, and drain-pipes, and bottles of various descriptions and capacities; but the tenancy of

UNLOADING A KILN.

these "narrow cells" here is reduced to eight days, I subsequently discovered, and when the kiln is full—and the contents may be numbered by the hundred—they brick up the opening and draw the chains over it and set the fires going.

The superfluous heat is utilised at the top of the kiln, which is on a level with the upper floor, for drying purposes. To put an article in to bake with the slightest suspicion of moisture in it would be disastrous, and a considerable plant in the shape of steam-heated pipes and chambers has to be maintained in addition to the drying spaces the kiln-tops afford.

Passing into the studios again, a gentleman, whose name I did not catch, kindly unwrapped the portions of a beautifully executed colossal Roman group, designed for the exterior of a large building, that he was modelling in clay. To be vulgar—a relief now and again—I was "flummuxed." And yet it all seem so delight-fully easy !

I got a piece of plastic clay myself when I reached home that night and designed a bust of my artistic companion out of it. I asked my wife if she recognised it, and she said, Yes—it was a ginger-beer bottle. She also thought it was "cute," and seemed to imply that such a proof of my capability was vastly creditable to me and encouraging to herself. She apologised, of course, when I told her who it was meant for —she said the tall hat rather misled her: she had mistaken it for the cork; but she seemed rather to sympathise with my model, and suggested that perhaps, after all, there were one or two details about such work that would be worth acquiring before doing anything for public exhibition. She said the libel laws were rather stringent, and it might possibly save worry and expense.

From the studio we wandered down through Pottery-Land, and crossed the road to a magnificent building in which terra-cotta work and ornamental tiling play a large part in the construction.

Here I beheld many quaint, curious, and costly things—the product of this land—that filled me with wonder and delight, and a vaguely envious feeling. Beautiful vases, richly-coloured tiles, and handsome fireplaces were on view, and the various elaborate sanitary and electrical appliances that are invented,

patented, and produced by Messrs. Doulton **and**
Co., showed in contrast, as evidence of the
magnitude and scope of the work done in
Pottery-Land, to.timepieces and Scriptural life-
size groups marvellously executed in terra-cotta.
A greater contrast still was afforded in a big
storeyard adjacent, where many miles of drain-
pipes are stacked to fill an order for the
sweetening of some Colonial town that **has**
recently awakened to the advantages of these
productions over the old-fashioned, dirty, and
dangerous brickwork contrivances that Messrs.
Doulton were solely instrumental in dispelling
for ever from our own cities, and that. not so
very long ago, either.

And so I turned and came away from Pottery-
Land, a land so important, so vast, and so
elaborately simple and simply elaborate, that I
know of none that can contrast with it in point
of interest and brain befuddlement.

IN MATCH-LAND.

IN visiting Match-Land I went by way of Victoria Park, which is in the East. The sun was shining and so were my boots, the birds carolled enthusiastically on the trees, and all nature was as gay as it could be with the wind also in the East and influenza raging. I turned out of the big iron gates into the Old Ford Road, and passed by mean dwellings and crowded and tumble-down-looking neighbourhoods; and when I at last crossed the canal bridge and entered the premises of Messrs. J. Palmer and Son, the resemblance between my boots and the sun had faded.

Glancing casually round at the land I had come to explore, I noted that it was built about on three sides, and that huge gates occupied the fourth and kept Match-Land private and select from the outside world. I noticed also the light bridge that runs across the court and connects the buildings, and the huge stacks of timber waiting to be converted into boxes and "splints"; and then I entered the offices, and to me, shortly afterwards, came Mr. Mountain, the manager, at whose kindly and courteous invitation I had journeyed down to view the mysteries of match-making.

"Begin at the beginning, please, and don't be technical," said I. I have repeated this so often of late that it is becoming stereotyped, and I use it unwittingly when I am not holding my attention closely to the matter in hand. Only the other day I went into the Stores, and a very civil and attentive young gentleman came up to me and rubbed his hands together and soliloquised with one of those loud sort of soliloquies.

"Good morning, sir," he said. "A lovely morning for this time of the year." It was raining in sheets, but I was thoughtful and agreed that it was a beautiful morning.

"And what can I have the pleasure of showing you, sir?" he went on.

"Oh, just begin at the beginning, please," I replied, pleasantly, "and don't be technical. What I want is——" I stopped in some confusion, and begged his pardon; but it just shows you how these things grow on one.

Mr. Mountain caught my wishes readily, and in a business-like way he led me forth from the offices and down the yard, informing me in transit that Match-Land receives its supplies from the canal below, and from the railway depôt lying adjacent. The same handy means bears away the manufactured product.

"Canadian spruce and pine," he remarked, briefly, as we passed by the huge baulks of timber waiting to be carried into the saw-mills; and I gathered that the match splints are cut from the pine, and the spruce is required for making the boxes.

Turning into the store-rooms first, I found in a series of bins and tubs and sacks the materials required in the works. I didn't catalogue them; but chlorate of potash, glue, saltpetre, and stearin were variously pointed out to me as being necessary for the proper construction of matches, along with reels of cotton—not the common ha'penny reels of commerce, but great spools holding miles of the soft and thick variety for the making of wax matches, and cases of phosphorus, that comes in in sticks packed in water in pound tins, soldered down and air-tight. The uses of these things were explained to me; but it will be clearer if I deal with them later when they are being utilised. The inhabitants of Match-Land find that that is the best occasion for getting satisfactory results, and a well-established precedent like that is not to be lightly forsaken.

The melancholy screech of circular-saws cutting through heavy timber greeted me even before we entered the mill; and, following my guide closely, I beheld a number at work reducing massive baulks into oblong blocks of a suitable and handy size for subsequent operations. As I before explained, Canadian spruce and Canadian pine are used. They get it from Canada. The pine blocks, after being cut transversely by a four-bladed saw into oblong sheets the thickness of an ordinary match, are

HORIZONTAL BOX-MAKING MACHINE.

fed into a sort of guillotine that quickly covers
the table in front of it with splints, each splint

being the length of two matches. It keeps one man pretty continuously at work making up the splints into bundles, which are subsequently put into a frame and carried into a gloomy cavern, that struck me as unpleasantly warm when I entered it, to be dried. The public like them better that way, I gathered, as being more serviceable and effective; and in small matters like these Match-Land doesn't mind stretching a point, even to the extent of maintaining steam-heated rooms for the purpose of obliging its customers.

In following the box-making process, I was led to a marvellously simple and a simply marvellous machine, into which the blocks of spruce, having been previously steamed to make them pliant and less liable to break, are brought under a knife which pares them into thin slices and sends them out with four lines scored across the surface at proportionate distances. My guide took one, and, bending it rapidly round, handed it to me, the rough semblance of the outside of a match-box.

.The illustration gives you an idea of the horizontal machine, and the principle of the

others is the same, only they work vertically.
The inside cases are cut and scored from
narrower strips in a similar way. The bottoms,
of course, do not require scoring, they are
simply shaved from blocks cut exactly to the
size required.

Each box, it will be seen then, consists of
three pieces—the outer case, the sides of the
tray, and the bottom—and these are given out
in sets, along with the paper that unites them,
to be made at home, in gross bundles. Ten
gross at a time is an ordinary supply, and where
whole families work at the box-making, as
many as fifty gross at once will be taken away
and made up. The inquiring mind may easily
familiarise itself with the method of scoring
and making these boxes by removing the paper
from one and spreading it out on the table. It
will be just as apparent if it is spread out on
the kitchen dresser, only I mentioned the table
as being handier.

I lingered long in the box-making room, in-
haling the scent of the freshly-cut wood, and
pottering about asking questions and laying in
a stock of mechanical information, while the
melancholy screech of the saws went on around

me, and the sawdust gathered upon my togger
—I mean wardrobe. Blocks and splints were
stacked about the building, and disputing the
floor space with spluttering little engines and
the machines they were operating. In one
corner I came across an exaggerated kind
of lemon-squeezer, and found that it was rather
a primitive sort of scoring-machine that doubtless did good and useful service in its time. On
its lower surface, which is screwed to a bench,
four blunt knives are set across, and the slip
being placed on these, and the upper half
brought down with a smart tap, the result is
obvious. Primitive as it looks, however, some
skill is required to manipulate it. The first
time I essayed to work it the scoring was not
perceptible; the second time it was obtrusively
apparent—it cut the slip into pieces.

Recognising at length that the " tempus "
was " fugiting " at an alarming rate, I turned
from the sawmills, and, crossing the court with
my guide, entered the buildings in which the
business end of the matches is made and
applied. The splints I had seen constructed
across the way do not serve for wax matches,
which probably will not surprise you, and the

taper to supply their place is also prepared here. I visited that department first.

STEAM CUTTER IN WAX MATCH ROOM.

Along one side of a spacious room two immense drums are standing, and in between

13

them is a shallow, enamelled, steam-heated bath, containing a liquid compound of stearin, paraffin, wax, and gum. Over one of the drums, which are operated by means of cog-wheels attached to spindles running through the wall, the cotton is wound up to form the tapers, each taper comprising from twenty to twenty-eight strands. The ends are led through the bath, and attached to the opposite drum, which is then set revolving. I described this process in "Candle-Land" when dealing with the taper-making department; but it is an ingenious operation, and pretty to watch, and will bear repetition. On coming out of the bath each taper runs through holes drilled in a steel plate, which removes the superfluous material, and also serves to press and bind the strands comprising it firmly together. When the carrying drum is quite unwound, the process is reversed, and this is repeated four or five times. The taper is then firm, hard, white, and shiny, as you see it in an ordinary wax-match, and is reeled off on to smaller drums ready for the cutting machines.

These "cutters" are small and compact affairs, with a complicated motion, being

interesting to watch, and ingenious withal.
Into the back the ends of the tapers are fed in
a straight line from the reels. I didn't count
them, but I should judge that there must be
upwards of a hundred. The machines are
steam-driven, and are operated by women, who
sit before them with one foot on a controlling
brake.

The frames for receiving the matches are
simple yet ingenious contrivances, which tend
to make the subsequent operations wonderfully
expeditious and compact. As the illustration
shows, they are merely stout, oblong frames.
The sides are rounded, and the cross-laths,
which are constructed to slide up and down,
are gathered up in the operative's hand, when
she introduces a frame into the machine. The
laths are lined with felt, and when the brake is
released the machine throws the row of tapers
forwards with one movement, cuts off the
required lengths in the second, after the
operative has released a lath which drops down
on to them and holds them in their places, the
whole frame dropping level for the next filling
in the third. And so the process goes on, a
lath being dropped after each cutting.

· There are about sixty laths to a frame, which
will thus contain when full between five and
six thousand matches ready to have their
business ends attached; and when the last fill
has been made, the top bar is fitted on and
screwed down tight by a couple of thumb-
screws, after the manner of the primitive and
patent trouser-stretcher, to hold the contents
securely in place, an operation in which the
felt linings materially assist. Female labour is
extensively used in Match-Land, and these
machines are worked by women, who will fill
as many as two hundred and fifty frames in a
day.

Three or four hand-machines were also at work
filling frames with considerably more noise and
bustle than their steam rivals. They filled
them with matches as well, and the process is
identically the same, except in the matter of
driving power. My guide called them "hand-
machines," though, as a matter of precision,
they are worked by treadle on the sewing-
machine principle.

The frames, being filled, are carried away to
be dipped, and before taking up this process I
may as well show you how the wooden matches,

either common or safeties, arrive at the same satisfactory stage, although the work is carried on in another building.

The splints, cut and made into bundles as I have explained, are dipped into a bath of melted paraffin and resin, and are then taken to filling machines bigger and stronger than those employed in the wax-match making, and with

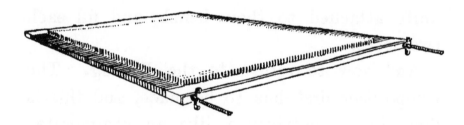

FRAME FILLED WITH WAX MATCHES.

deep troughs set at the top, into which the supplies of splints are fed in lieu of being unwound from reels at the back. Wooden splints wouldn't unwind satisfactorily, so they have to vary the programme somewhat to meet the prejudice these inanimate objects so frequently display and persist in. They are shaken down the groove, and, finding their level, are thrust forward into the frames by a row of metal bars, which, moving automatically, push them well forward into the frames. The laths grasp them in the centre instead of at the end, and present even rows of gleaming white pegs on both sides. There is a beautiful economy in this—the economy of labour and time which serves to make two matches in the place of one; for, being dipped on each side, the fillers gather them into a bundle and sever them in the centre by a lever knife attached to their bench, and fill each two boxes.

And now to return to the dipping. The composition first has to be made, and this is done in a contrivance like an exaggerated saucepan. For the ordinary match composition, glue is melted up with chlorate of potash, phosphorus, and dead matter—mostly ground glass. There is taste in matches, as well as in everything else, and to satisfy the artistic prejudices of the customers aniline dye must be added. Personally, I am impartial in this matter, and would as soon light my pipe with a blue match as a green one; but others are more particular, I was somewhat surprised to learn, and blues, browns, and reds are specially prepared to meet every particular fancy. The dead matter is added to make bulk, and also to put a brake on the active lighting principle, the potash and phosphorus, which, without its gentle restraint, would flare and exhaust itself when ignited without catching the stick. The glue holds the mass together, and also enables the composition to attach itself quickly and firmly in the dipping process.

They were stirring in phosphorus when I appeared on the scene, and out of a tin just opened—I think I explained that it comes into Match-Land in rounded sticks soldered up in water in ten-pound tins—my guide handed me a sample.

"If the room was warmer," he remarked, casually, as I turned it over and smelt at it, "it would flare away in your hand."

I hastened to replace it then, and went to look at something else. They might be wanting to use it, I reflected; and besides, our climate is very uncertain and changeable.

The composition being duly mixed and stirred up to a proper consistency is taken away to the dipping-rooms. A smooth iron bed-plate, which is kept heated by water and a gas jet underneath, represents the stone in the

sort of lithographing process that is adopted to
fit the matches for a useful as well as an
ornamental career. The composition which is
spread over it—the table I mean, not the career
—may be similarly likened to the printer's ink.
It is smoothed out by a knife arrangement to a
thin, even surface, and it is curious to note how
the particles that become detached and dry will
snap and flare when the blade touches them.
The workman then takes a frame in hand, and
lays it flat down on the composition, and when
he removes it again the matches are finished.
In the case of the splints, the frames are left
for a few minutes for the composition to set,
when they are turned over and similarly treated
on the other side.

From the dipping-room the frames are taken
away and lodged in racks in a series of
chambers that would offer the chronic weather
grumbler just *the* thing he requires in the way
of an adjustable climate. They are fitted with
steam-pipes and immense steam-driven fans, and
the temperature can be, and has to be, regulated
in opposition to the weather that maintains
without. Damp spoils matches, you will be
interested to learn, and so does undue heat. By

DIPPING.

means of fans and pipes they can blow hot and cold at the same time in these rooms, which is considered rather a shady thing in a man, but is excusable in a drying-room.

It was pleasantly cool when I went into one

after the heat of the dipping-room; but I gathered that it could be made unpleasantly warm, if occasion required, with equal facility.

After being dried, the frames, in the case of wax matches, are taken into a long and lofty room, with long benches fitted up and down the centre and round the walls. Before these benches girls and women stand working unceasingly, filling the boxes which lie in a heap by their side. They work very rapidly, the matches, loosened in the frames previously, being taken out and boxed in two movements. They are filled in with the heads pointing both ways to make them set evenly, otherwise they would bunch up and you wouldn't get enough for your money, and a long tin blade is laid over them and the cover is slipped on. This facilitates operations, and saves time and loss by accidental ignition. Even as it was, an occasional box would flare up during the filling, and be set on one side to burn itself out, and be counted in with the bad debts of the firm.

I never saw such a place for matches. You brush them off the benches with your coat as you turn round, and they explode unexpectedly under your feet and make you jump as you walk about the room.

The boxes vary greatly in size and shape, and range from little pill-box-shaped affairs that are very popular in the Australian market, to broad and lengthy tin cases that almost require a hand-bag to transport them. Many of them are boxed for special customers, who have their own labels affixed for advertising and other purposes that are sufficiently obvious.

And then the boxes are made into dozen parcels, and are taken away to the packing rooms for the last stage of all. Being perishable articles, the cases for export have to be metal-lined, and the contents soldered down to withstand the friendly overtures of the sad salt waves and damp atmospheres.

It was while viewing the packing operations that I discovered the mystery of the safety match, and the reason for its exclusiveness in the matter of only igniting on the box. It is this. The igniting principle of matches is the admixture of phosphorus and chlorate of potash, as I have already explained. Well, they make up the composition without the former ingredient, and incorporate it instead in the black strip outside the boxes. There is no fraud in this, though it is a bit artful, because they give you the box and the contents for the one cost; and if you like to carry your matches loose in one pocket and leave the box at home behind the door in another, that is, of course, your matter, and one that the proprietors of Match-Land cannot reasonably be expected to hold themselves responsible for.

In the next department I came upon a very interesting operation, and withal a melancholy spectacle. The manufacture of braided-lights was the operation, and dozens of machines standing idle in rust and dust and decay contributed the dismal element. A decaying industry is always a melancholy spectacle, and the match-makers have much to complain of in this direction. In the case of the braided-lights—at one time a very popular article with smokers, but now neglected and despised, and dubbed "stinkers" in society that is not polite—prejudice has turned against them and they are considered bad form, so that where one box is used now-

IN THE DRYING ROOM.

adays a hundred would have been required in
years that have not long gone by.

The importation of cheap foreign matches, too, is dead against the well-being of Match-Land. From what I had seen I could readily understand that a retail price of three-halfpence a dozen puts a very serious obstacle in the path of the British producers. My guide spoke mildly and with gentle forbearance on the subject; but I was indignant, and just for once I trotted out my latent stock of patriotic fervour and gave it an airing.

"It is scandalous," I remarked, sweeping away Mr. Mountain's attempts at pacific explanation and a pile of matches standing on the table near by with a gesture of indignation as the monstrous nature of the whole thing spread itself out before me. "We are being ruined by Continental cheap labour and the Government wrangle over Ireland and music-hall promenades and such-like things, while the British workman is being forced into the gutter and the steak pudding is taken out of his mouth and the bread of charity is substituted in minute wedges in its place."

In my excited contemplation of this crying injustice I raised my temperature to a some-what uncomfortable degree and, in pulling out a handkerchief to mop my brow, chanced to dislodge with it a box of matches that I usually carry with me in case of fogs. It eluded my grasp and rolled under the feet of my guide, who picked it up and handed it back to me with a smile that was half amused and half thoughtful. The box said they were "made in Flanders," and arranged its position so that all who walked by slowly should know it.

"It is not altogether the fault of the Government," he remarked, briefly, and I, losing my interest in political economy, hastened to inspect the processes going on around me.

Braided-lights are familiar to every one, with their rounded sticks and plaiting of gaily-coloured cotton and exaggerated tops. The thin circular rods come over in bundles from Germany in forty-inch white, pliant, and shapely lengths. At a row of plaiting machines, which work with a terrific rattle and noise, a number of women stand and feed them one by one upwards through a hole drilled in the steel bed-plate.

The bobbins that you see are filled with the

thin yellow cotton from big reels by a machine standing at the back of the "plaiters," and the wire that runs up the sides of the stems are similarly wound on small bobbins and are set underneath the machine. It is a continuous process, the various strands being connected to a new supply as each bobbin becomes exhausted while the rod is automatically lifted as the plaiting progresses, and is followed closely by another before it leaves the machine to be coiled up on a huge drum above.

"Why are they wired?" I ventured to inquire of my guide, and by the exercise of much mental effort the observant reader will detect a subtle jokelet in the question.

"To hold the head in its place while burning," was the reply, and being satisfied thereat, I watched the rods in one continuous length of wood, wire, and cotton being passed through a solution of gum to keep the plaiting from unravelling when they are divided into their proper sections again. They are then cut into two-match lengths—the same as the ordinary variety—and are dipped in three processes. They are dipped in composition as well—the first two in a mixture of potash, saltpetre,

PLAITING IN BRAIDED-LIGHTS ROOM

charcoal, and some peculiar incense that con-
tributes the "smelly" element so noticeable

14

when they are ignited in a close railway-carriage, and the third in phosphorus, which completes the operation. It likewise concluded my observations as far as this particular industry was concerned, so, thanking my guide for his kind attention, I gathered up my over-coat and umbrella, and wandered forth into the mud.

IN RUBBER-LAND.

I FOUND Rubber-Land at the close of one bright but chilly afternoon in March—though, of course, it may have been there before then—in a quiet street turning out of the Burdett Road, and situated within a few minutes' walk of the station on the Great Eastern Railway bearing that name. Candidly, I was disappointed with its appearance at the outset. There seemed to be nothing of promise contained in the long low block of buildings, with the staring rows of flat windows

and the dingy colouring that had once been
white, with the inscription,

ABBOTT, ANDERSON, AND ABBOTT,

in huge letters showing blackly against it, to
give indication of the plant and premises
extending backwards in a square block and
built in on every side with houses and manufacturing premises.

Dod Street, Limehouse, is certainly not a
place for the poet to dwell upon, and it didn't
strike me as a locality at all suitable for this
particular scribe to dwell in ; for mere exploring
purposes, however, I resolved to let it serve, and
went in to Rubber-Land by a side door to investigate its mysteries, and to shed the light of my
presence over the inhabitants, just by way of a
treat for them.

I am glad that my better nature triumphed
in this matter, because they were so genuinely
pleased to see me, and so courteously attentive,
that I would have gone into a backyard and
praised the dust-bin and admired the water-
barrel just to reciprocate. But they really had
something to show me, and by the time I
reached the street again I had no feelings but

those of the completest gratification and
admiration, save, perhaps, a stray pang of
hunger that betokened the passing of my usual
hour for tea.

Mr. Jackson generously took upon his
shoulders the onerous task of guiding me
round, and playing the limelight upon my
ignorance regarding the processes by which
rubber and cloth are amalgamated into a
damp-defying article.

As he steered me past packing-tables and
huge bales of cloth and finished garments, he
made clear to me the nature and extent of
Messrs. Abbott, Anderson, and Abbott's operations, which are about equally divided between
every and any variety of mackintosh article and
oilskins. The last comprises sailors' and
fishermen's garments against the overtures of
the elements "when the stormy winds do
blow-ow-ow," and policemen's capes and
leggings, and 'bus-drivers' aprons, the firm
having standing contracts with the City and
Metropolitan Police authorities, and several of
the 'bus and tramway companies for the supply
of the latter items. "No," said my guide, in
answer to my query, "we do not do much in

the way of manufacturing mechanical articles. Tubing? Oh, yes, we make tubing and water-bottles, and water-beds and that kind of thing; but oilskins and mackintoshes are our specialities."

"Lead on, McDuff, I follow," said I, grasping my pencil, with the light of grim determination in my eyes.

"I beg your pardon," said my guide, politely, turning.

"It is granted," I observed. "Begin at the beginning, please, and don't be technical."

My ideal factory is plenty of noise, plenty of smell, and plenty of wheels; and two minutes later I realised it, which is a thing we are popularly supposed never to be able to do.

The rubber comes into Rubber-Land direct from the docks, which are conveniently situated barely half-a-mile away, in the shape of Dutch cheeses, packed in big oblong cases. It is curious and most unpromising looking stuff, and is obtained from Africa and Brazil. I expect that is why they call it indiarubber. That obtained from Para, I gathered, is the best quality of rubber in use.

The natives give it the Dutch cheese form by

a continual dipping process that builds it up in thin layers, and it is possible to follow the process by obtaining a section and separating these layers, which spread out over the fingers like a thin, transparent skin.

The crude rubber is stored away as it is received in big underground cellars until required for use.

The first process consists in boiling it in a large, open, steam-heated tank for several hours, in order to soften it, after which it goes through a set of horizontal rollers, with scored surfaces, and such a terrible crushing power that the contemplation of them at work made me shiver. The rubber is too hot to comfortably hold after it has passed through them and emerged underneath, no longer cheese-shaped, but in a long, thin sheet, despite the-fact that taps overhead pour a continuous supply of cold water upon it during the operation to keep the temperature down, and also to wash it—the rubber I mean, not the temperature. I thought the water was hot at first, seeing the steam rising in a cloud above the machine, and I covertly put my hand under it, to see if they hadn't made some mistake, even after I had been

assured that it was not. But there was no
deception, it was cold. It ran up my arm and
wetted my undergarments, and impressed the
fact upon me for the remainder of the
afternoon.

The men stand by the machine and gather
the sheet up, and break it off in sections and
throw it back between the rollers until it has
been thoroughly cleansed, and it is astonishing
what a lot it makes of itself under this treatment. It seemed as though it would go on rolling
out yard after yard of sheet from one little supply, and swarm over the room, and gather us all
up in its warm and moist embrace, and bury
us out of sight, so that they would have to come
and dig us out with pickaxes and dynamite, and
bury us cheaply and unostentatiously out behind
the naphtha-house, with perhaps a stained-glass
window to commemorate the tragedy, in the
frame overlooking the yard, and a tablet underneath inscribed with all our names, and the fact
that "here perished the only original explorer
out of Africa, and his trusty guide, victims to a
compound of duty and indiarubber."

Having got it into nicely-patterned sheets, in
the innocence of my heart I imagined that there

was nothing left to be done but to fix it on to the cloth and send it out.

CRUSHING MACHINE.

I made some such statement to Mr. Jackson, who laughed frankly.

"How would you fasten it on?" he
queried.

"Tin-tack it, perhaps," I hazarded.

"Why not gum and safety pins?" he
retorted.

As a matter of fact, I discovered, so far from
being finished, the treatment had hardly commenced. The three sets of horizontal rollers
through which the rubber has been passed and
repassed have brought it into a condition to be
taken into a steam-heated chamber—a stuffy
and smelly apartment, heated up to ninety
degrees—where the long, black, and leathery-
looking sheets are slung over rods near the roof,
and left to dry.

This being satisfactorily accomplished, they
are taken down and away to a curious machine
termed a "masticator." This is an iron box of
immense strength, set horizontally, and having
a curved ribbed roller revolving within, with
only the space of a few inches between the
inner walls and the roller. An oblong opening
on one side serves to admit the material, and
this is covered, when the "masticator" is fully
charged, by a grating of thick steel bars, made
in three sections, which drop down and are

primitively and securely locked in their place
by metal wedges.

At first the traps stand open, and the rubber
sheets are doubled up and thrown in, where
they are caught up by the roller and carried
round slowly, and forced into an oblong ball
again. Then as the supply is added to, first
one trap is closed and then another, and, with
a pressure of many tons upon it, the huge mass,
creaking and groaning and smoking hot from
the friction, is carried round and round for three
hours, by which time it is thoroughly kneaded
together in one hard, solid, oval ball.

Rubber is a stubborn substance, and even
this treatment does not quite suffice to bend it
to the will of those who have its manipulation
under control, so they take it in hand once
more, and place it between a pair of horizontal
rollers, similar to those in which the first
operations were effected, except that they are
made of chilled steel, and are smooth and
highly polished on the surface. These rollers
all work towards each other, so that the operatives have merely to feed the supply in between
them, and to gather the sheets from underneath as they emerge, and throw them back

until they are in fit condition for further treatment.

Up to now the rubber is a solid substance, and all this treatment has been necessary in order to soften and prepare it in thin sheets to readily absorb the naphtha that is to convert it into what my guide termed " dough," and what I mentally dubbed a sticky-looking mess.

Naphtha is a dangerous commodity, and, in consequence, the wise old grandmothers who so carefully attend to our morals and our manners, our lives and our liberties, have enacted that any process involving its use must not be carried on within so many feet of the main factory. Accordingly we had a pleasant little stroll around to the unpretentious building where the rubber sheets are steeped in a number of shallow metal cans, in the process known in Rubber-Land as " soaking down."

Naphtha and rubber are put into a bath, in due and just proportions, which is covered with a closely-fitting lid. Ever and anon a man will come and stir the gluey-looking mess about with a pitchfork affair, just to see how it is getting on. Then there comes a time when they find that it has " got on " sufficiently for their needs, and the "dough" is ready for use. It does resemble dough in its consistency,

THE MASTICATOR.

though it is of a deep black colour and redolent of naphtha and warmed-up goloshes.

There is a comic element in its subsequent
treatment that appealed to me. It called up a
suggestive sling-me-and-I'll-stick-tighter-than-
wax sort of feeling, that impelled me to wish
that I had some object of private detestation
around, that I might gather me a handful and
heave it at him. It is very wrong, I know, but
some things do affect you like that. The
gentlemen and ladies down Whitechapel and
Shadwell way experience the same sort of feeling with peas-pudding; they can't help it; it
is something in their temperaments against
which it is useless for them to strive, and the
pathetically hopeless look they will cast at you,
along with a liberal supply of the dainty compound, should you happen to be strolling their
way in a tall hat and a decent coat, has a
tendency to make you feel wretched. It is
chiefly on that account that I never go their
way on peas-pudding nights, which are usually
Mondays and Fridays. Appropriately enough,
the " dough " is treated in " dough machines,"
a series of smooth steel rollers that revolve in
opposite directions, and pass under metal knife-
shaped blades, that are adjusted by handscrews
above them. Working the " dough " is cer-

tainly a messy operation, since the men have
to handle it continually, though a pail of water
set at hand, into which they ever and anon dip
their fingers, prevents its loving attachment
becoming too embarrassing.

The "dough" is removed from the cans in
which it is brought from the naphtha-house, and
is laid between the rollers. Adhering to these,
it is carried round, and scraped off by the knife
previously mentioned. There it accumulates,
until the men come round, and scoop it away
with their fingers, and either throw it back
between the rollers or into a metal can ready
for its reception. This operation is to absorb
the naphtha, as I gathered from Mr. Davis, the
foreman, whose kindness in spending a great
portion of his time in making me familiar with
the technicalities of the machines and rubber-
treating processes I must in justice acknow-
ledge. Surely never was explorer so thoroughly
served before! He and my guide took me in
hand between them, and hammered the facts
into me, thus leaving me nothing to do but to
add a little colouring, to make them pretty—
but which will involve you in no extra expense
—and give them out again.

A similar set of rollers, working on the same principle, is used to incorporate the colouring matter, which is put in dry with the " dough," and thoroughly mixed up in the operation.

There were metal-lined stock-boxes of coloured "dough" standing about, all closely covered, and a removal of the lids revealed to me the æsthetic bliss wrapped up in the fact of being able to obtain a mackintosh rubbered to suit the complexion, blue, brown, white, or just a common black.

Whatever the colour, the " dough " is then transferred from the stock-boxes to a vertical drum, into the bottom of which a metal perforated plate is fitted, to form a sort of a colander, and the contents are squeezed through it forcibly by a piston-rod, actuated by a screw and wheel, into cans set underneath. And in these cans it is transferred to the regions above, to be practically utilised in the manner that I will presently detail.

Before leaving this building, I may mention that a comparatively small proportion of rubber is steeped to a thinner consistency than the "dough," and is subsequently churned in an ordinary steam dough-mixing machine—a square

A MIXING MACHINE.

box with four curved blades revolving in reverse
directions within. It comes out as much like

15

treacle as anything, and is used for sticking the garments together—and great is the "stick" thereof. Fairly put on and pressed, you can as readily tear the stuff as drag the united surfaces apart when joined up by this means, and it has other obvious advantages over machine or hand stitching. There is, in fact, no sewing in mackintoshes except in the buttonholing.

"And now we'll go upstairs," remarked my guide, and upstairs we went, finding a convenient flight set outside the door, in the open, which was a handy arrangement as far as interior economy of space was concerned, though it would have its drawbacks for ordinary domestic use. I went up twice. I went up the first time because I wanted to get to the top and see what there was to see, and I went up the second time because, in my haste and anxiety, I dropped my notebook over the balustrade, and had to go down again to recover it.

I got up, bag and baggage, eventually, and stood in a spacious chamber, the floor of which was mainly occupied by four broad metal tables; and it was here, I gathered, that the cloth is brought to have its damp-resisting qualities added. The smearing process is still adhered to in "proofing" the cloth, and indeed, the gluey rubber renders that treatment most natural.

The tables are oblong-shaped contrivances, and are heated from below by steam pipes to a temperature that would not induce you, should you inadvertently perch yourself on one to rest and meditate, to linger there any longer than necessary. Indeed, on the bleak afternoon that I made my inspection, the room itself was so hot and oppressive with the fumes of the naphtha that I was jolly glad to get outside again.

At the head of the table an arrangement of rollers is seen, as the illustration shows. The cloth, which is wound on to spindles for treatment, is set in the front rather below the bed, and is led away through big rubber-faced rollers, over the table and back underneath, where another spindle gathers it up again.

Just behind the operatives, zinc-lined stock-boxes are fitted to contain the "dough," which is maintained in the soft, gluey consistency in which it comes up by wet rags that prevent the naphtha evaporating. A knife, similar to those on the dough-making machines down-stairs, is set over the rollers, and the distance between is regulated by hand-screws, so that they can obtain any thickness of rubber on the cloth that may be deemed necessary. Two adjustable pieces of metal are set at the sides in the same way, to keep the rubber from spreading. A liberal supply of "dough" is then taken from the chest, and smeared along the cloth at the base of the knife, and the machine is started.

The inexperienced "dough" chuckles at first to think what a time it is going to have. It sees all sorts of possibilities of rambling round that table, and messing up the floor, and clogging the wheels, and spoiling the men's work. Its mirth is of short duration, however, for immediately the cloth begins to travel it is carried up against the knife, and the portion that does manage to get through covers the cloth with a thin, even sheet of rubber. The heated table evaporates the naphtha during the journey across it, and when the end of the roll is reached the knife is lifted and the cloth is wound back and the process is repeated over again. The number of coatings given is regulated by the service the cloth is to be put to, the ordinary mackintosh cloth requiring six.

PROOFING THE CLOTH.

Precisely how many treatments shall be given is determined by weight, there being a regulation standard for the various classes of goods.

If made up into garments without any further treatment, the rubber would rot with the action of the weather, so a sort of setting process has to be gone through, which is termed "vulcanising." This is effected by heat, or by sulphur, and to witness these operations, I left the building, and traversed a light staging that runs round on a level with it to an adjoining block.

Dante's Inferno was suggested to me while making my inspection, that is, if you can imagine a sort of domestic Inferno constructed out of connected chambers, with heavy, lead-lined double doors, which meet in the centre and are screwed up from outside. On the floor a number of pans for burning the sulphur in are dotted about, each with its gas jet under it supplied by rubber tubing from pipes down the sides. The cloth is suspended from bars set across the top, and when the chamber is full it becomes somebody's unpleasant lot to set light to the gas and get out without any unnecessary delay. There seemed so much possibility for a few first-class tragedies here that I experienced quite a thrill of anticipatory horror as I remarked, carelessly, so as not to allow my eagerness for sensational "copy" to become too apparent: "I suppose you very frequently lose a man or two in there?"

"Oh, no!" rejoined my guide, cheerfully.

"Not one?" I asked, disappointedly, feeling that I had been defrauded somehow in the matter-of-fact denial. "Not even a little one —a boy, perhaps?"

Mr. Jackson shook his head. He evidently sympathised with my disappointment; but, with George Washington's record before him, he could not strain a point, even to accommodate me.

"Ah, well," I sighed, resignedly, resolving to let the tragedy go; "but I suppose there have been many narrow escapes—overcome with the fumes—man rushes in—daring rescue —and all that sort of thing?"

"No; oh, no; not at all!" said my guide, conclusively, and I had to give way. I was more than disappointed, though; I was vexed; there seemed enough possibilities in the thing to construct an Adelphi melodrama from.

In an adjoining chamber I witnessed the "dry-cure" method of vulcanising being

carried on in a huge wooden, metal-lined
compartment that is filled with rollers and
heated by steam to two hundred and fifty
degrees. A miniature engine works these
rollers, and the cloth is let through a long,
narrow slit from the rollers outside, and sent
on a devious journey that occupies an hour and
a-half before it emerges again overhead to be
wound up on another spindle. Here, too, a
man has occasionally to crawl in to repair a
breakdown. The idea of groping one's way
into that dismal box among hundreds of yards
of cloth in such a temperature, heavily charged
with sulphurous fumes into the bargain, gave
me the horrors; but there were no casualties to
report.

"What does it suggest?" I queried of the
man in charge, hoping to draw him.

"It suggests getting out again as soon as
possible, sir," he replied, laconically. It is a
reasonable suggestion, too, when you come to
think of it.

After this the cloth is taken away and
doubled. Two lengths of cloth meet together
between heavy steel rollers, and the rubber surfaces unite of their own accord, assisted thereto

by the persuasive pressure that is put upon them. And then it goes up to the cutting and

THE TUBING MACHINE.

making departments, where feminine labour transforms it into mackintoshes, and capes, and other dress items so serviceable in our moist and uncertain climate.

I spent half-an-hour or so in the making-up rooms, in which dexterity of hand-work was apparent, and the rattle and noise of countless machines was unpleasantly obtrusive. I also explored the oilskin buildings; but there was nothing much to record. It is smelly and very messy, the garments being made of cotton or linen, rubbed over continuously with oil until sufficient is absorbed, and again when they are made up. They are then sweated in heated chambers, and are varnished, and ready for the road, the river, or the sea, according to their nature.

"It is very messy and rather unsavoury work," said I to my guide.

"It is well paid," said he.

"It ought to be," said I. "I should want something considerable myself before I consented to veil my identity in oil, so that my friends and acquaintances would be unable to distinguish which was me and which was linseed."

One curious little construction I passed, on my way out of Rubber-Land, in the shape of a

sausage-machine sort of affair, in which rubber
tubing is made. Dry rubber, worked soft, is
introduced on top, and the tubing flows out
at the end in a continuous coil that, like
Tennyson's famous brook, flows on for ever, as
long as material is given it.

The shades of eve had done falling when I
reached this last stage of my explorings, and
the stars and the gas lamps were twinkling in
their respective twinkling places, so I thanked
Mr. Jackson for rendering my visit to Rubber-
Land so pleasant and profitable, and wended
my way to the railway station.

IN WIRE-LAND.

TAKE a ticket to the West India Dock Station," said Mr. Pelham Bullivant to me, on the memorable morning that I set out to explore Wire-Land, at the close of a brief interview during which I duly made my wishes known unto him. "There's a 'bus running from there which will set you down at our gates."

And I thanked him and withdrew; and I took the ticket and the 'bus—or, rather, the 'bus took me—and was shortly after come unto the Isle of Dogs.

Mr. Gilbert's famous admiral, who never ran a ship ashore, evidently doesn't live down Millwall way. There they seem to have been practising that kind of sport in their spare half-

hours, and the result is picturesque but not altogether pleasing. It lacks dignity, to my mind.
As a rule, you associate steamships with heav-
ing, white-crested waves and blustrous winds—
with the "i" accentuated, please—and the
poetic mind receives a shock when it sees an
imposing specimen of Neptune's playmates
endeavouring to poke out a coffee-shop pro-
prietor's bedroom window with her bowsprit,
over a tarred fence which separates her from
the traffic in the street, and prevents her from
climbing up over the roadway and getting
under the horses' feet and hindering them
from their lawful business. It is not in
keeping with the fitness of things, I think.
However, I had not travelled down purposely
to regenerate Millwall, so I held my peace
while the dirty little 'bus jolted me roughly
through long and dingy thoroughfares, and
carried me over bridges constructed above
lock-gates, until I was thankful when the time
came for me to deliver up my penny through a
hole in the roof, through which the driver's
hand protruded, and alight on the threshold
of Wire-Land.

I knew it for Wire-Land though the gates

bore the brass-lettered statement, " Bullivants,"
and I went inside and presented my credentials.

In due course a gentleman came round to me
and announced that he was Mr. Selby, and that
he was at my service.

I told him then my intention and purpose,
and added a fancy outline sketch concerning
the depth and extent of my remorse at so
victimising him.

He said he was pretty robust and calculated
that he should survive it, and seeing him take
the matter thus lightly I resolved to go ahead,
nor let any sentiment of pity hinder me from
doing my duty to the British public.

The operations in Wire-Land comprise the
manufacture of any and every variety of wire
ropes, wire netting, and torpedo - nets, and
appliances for working in conjunction with
them.

The ropes are used for manifold purposes,
and are of all sorts, shapes, and sizes, varying
from a pliable hemp-covered line which cannot
be readily distinguished from an ordinary rope
up to cables six or eight inches in diameter for
working tramways and lifts, and hawsers for
towing purposes.

In view of the fact that Messrs. Bullivant
and Co. are the largest manufacturers of their
kind in the world, I need hardly state that
Wire-Land covers an immense tract of ground
which stretches back from the main entrance
in West Ferry Road to the river, where
the company has a wharf of its own, with
steam-cranes and other appliances for handling
the weighty products of their manufacture.
Facilities are further afforded in this direction
by a complete system of trolley rails which
runs throughout the premises and passes into
the main buildings. The advantage of this
can be estimated when it is considered that a
steel cable weighing upwards of thirty tons is
not an unusual nor an outside order.

As most of the manufactured product is
shipped away from this wharf, so the bulk of
the raw material comes in by the same route,
being transferred from lighters on to trolleys
in the shape of huge coils of iron and steel
wire of various grades, which are first lodged
in a series of immense store-rooms.

It was to these that I was first conducted.
Tens of thousands of coils of wire were lying
around the floor space and stacked up in bins

round the walls just as it is received from the wire mills away off in distant Lancashire.

Hemp-rope is also used in Wire-Land, a fact which astonished me somewhat, to form the core of the ropes and cables manufactured.

My guide did his best to make me discriminate between one quality of wire and another; but it was wasted effort; my understanding was strained to the utmost to grasp and retain the fact that ropes could be made of such material at all. And then I was led into the rope-machine shop, a tremendously long building—one little machine in it that was packed over against the further wall measured three hundred and eighty feet—set over every square inch with the biggest and most marvellously complicated machinery that I have ever beheld. Hundreds of men and boys were at work there, and the rattle and clash was simply indescribable. Talking seemed out of the question; but nevertheless we managed to make ourselves understood, although I don't think this would have been possible if Mr. Selby hadn't had the particular machine we were inspecting stopped while he was explaining its action to me.

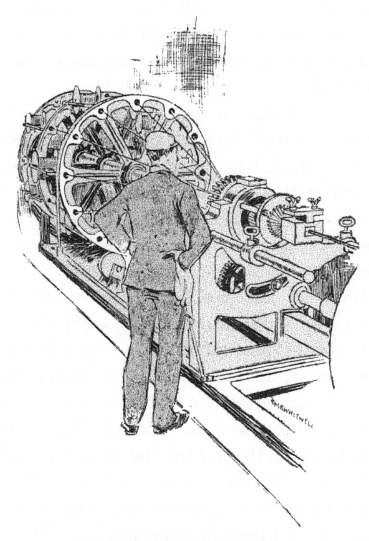

ROPEMAKING MACHINE.

Wire-rope making has many points in common with the manufacture of ordinary rope. The wire is first of all wound off on to bobbins

16

—big iron reels that will weigh when full from
three and four hundredweight to as many tons—
and are then taken to the stranding or closing
machines.

It seems that it is a vital point not to
"throw a turn" into, or twist, the wire when
working it, and in consequence it is necessary
to swing the bobbins on frames that will
always retain their horizontal position even
when being spun over at one hundred and
thirty revolutions a minute.

Most of the making machines are long
horizontal contrivances which carry from six
to twelve bobbins set round an open ironwork
cage affair, as the illustration shows better
than I can describe. The various strands
converge up near the head of the machine,
and through the centre the rope core is led.
When the machine is started, it lays the wire
round this core in a process very similar to that
which I described in rope-making. The rope
core is put in for the purpose of pliability, and
is previously steeped in oil to preserve it.

The ingenuity of these mechanical contrivances is something astonishing, and surprise
and admiration struggled within me and gave

me a headache—unless, perhaps, it was the
noise that gave me *that*—as I stood and watched
one monster whirling tons of metal round—
the bobbins alone would account for five—so
rapidly that I could only see a white gleam
where they were situated, though my guide
assured me that the operatives could follow
each bobbin, and be ready at a moment's notice
to stop the machine if a wire should break.
And the result of it all is the flexible, beautifully woven steel rope that is being coiled
slowly upon a big drum at the end of the
machine.

As in rope-making, the after treatment
is merely a duplication of process—strand
within strand to form rope, and rope within
rope to form a cable. Grasping this fact, I
strayed round and watched the machines and
the men at work, and my guide kept me attentive company.

"Cut a wire, there," he would observe to a
man in charge, and before I could protest the
machine would come to a halt and a wire be
parted. In the event of this happening either
purposely or by accident, the ends are cut off
clean by a pair of powerful nippers, and twisted

round each other, or tucked, or brazed together, and the machine is started slowly until the spliced portion has run through. The man in charge of the work knew whereabouts in the length of strand being formed that occurred, and so did my guide, but I didn't, though I felt and examined diligently.

I came at length to the 380ft. machine that I alluded to a little way back. It is a low, horizontal construction, and has the bobbins set up and down its length at regular intervals, the core coming off the last, counting from the foot. It was not working while I made my inspection, at which fact I was not unduly grieved ; but I could understand and follow its action by which the strands are led along and finally laid around the core, which in this case consists of an already manufactured stout steel rope. A big wheel at the head records the length of the cable as it is wound off, and three thousand fathoms is its average showing for a day's work.

Another fearful complication of wheels and work was shown me in a " closing " machine— an immense structure that towers high above you, and affords a walking tour in brief to examine thoroughly—which takes already decent-sized ropes and lays them up into one

CABLE COMING OFF THE MACHINE.

substantial cable weighing upwards of forty tons, and fit to do any service, from slinging

hundred-ton guns about to towing an Atlantic
liner around the globe—the wet part of course.
A recent exploit of this machine was the
making of a six-mile cable for the Melbourne
Tramway.

Unlike the hemp rope-making variety, it
must be remembered that these machines carry
their supplies with them, and a thirty-ton cable
means, of necessity, thirty tons of steel wire
alone, to say nothing of the ponderous frame-
work on which it is carried, whirling round
at a prodigious rate. But contrary to my
expectation,.the noise this monster made when
Mr. Selby set it in motion was easily surpassed
by its humbler prototypes, which showed
powers and possibilities of rattling a man like
myself not brought up to them into a lunatic
asylum any time within a month. There was
more draught than noise, and I am more than
half inclined to attribute a stiff neck that
I acquired that afternoon to its agency in
emptying half a gale of cold atmosphere down
my back when I wasn't looking.

The effect of the bobbins on these machines,
all waltzing sedately round like partners in
some curious outlandish dance, is very quaint,

and I was induced to untimely mirth on more than one occasion when witnessing it. Another contrivance that amused me was made in the shape of a scold's cage, or, for the benefit of those who are not acquainted with this old civilising influence, of a hooped skirt. The core here rises through the centre, and the bobbins are set round it.

The latest stranding machine, a handsome and complicated affair, stood away in a shop by itself and fairly filled one side of it. It measured ninety-eight feet in length, and could be worked separately as three machines, or, joined up, as one.

Before quitting this section of my exploring, I will endeavour to make you understand, as I understand it, the ordinary process of making wire rope.

In the first instance, the machines vary in size and capacity, and will carry anything from six to twenty-four bobbins. The process in any case is exactly the same; but the number of bobbins and, of course, the gauge of the wire determines the thickness of the rope. As the machines revolve, the wire is drawn from the bobbins through the head and twisted into a

"strand." Six or eight "strands" again laid
round a rope or metal core will form a rope,
and six of these ropes again plaited will make
a cable-laid rope. It is suggestive of the
weights and measures tables, isn't it? Of
course, I am using these figures for ordinary
requirements. There is a considerable lati-
tude in the actual performances; but the
big machine just mentioned is capable of
meeting them all. It will throw off a
cable in one length weighing forty tons, and
another having a circumference of twenty-one
inches.

Deeply interested as I was, I was not sorry
to at length turn away from the main rope-
making building, with its awful din and clatter,
and its grease—there is oil everywhere, and the
floor is black and slippery with it—and walk
out to the comparatively peaceful and sylvan
retreat along by the water's edge, where
there was nothing, save, perhaps, the conversation of the passing bargemen—who talked
to each other across intervening spaces of water
of a few hundred yards in that picturesque
dialect so peculiarly their own—and the hoarse
tooting of some big steamship, or the syren of

THE GALVANISING BATH.

a fussy little tug, to mar the silence of that rural spot.

I felt then, like Mr. Carroll's walrus, that

the time had come to talk of many things, and
during the interval I elicited some curious and
interesting facts from my guide. For instance,
unaided you would never come to believe that
the more wires a steel rope contains the greater
is its pliability, yet it is a fact, I assure you; I
have it from the most reliable source. Similarly,
torpedo nets, which are supplied in large
quantities to various Governments from Wire-
Land, and of which Messrs. Bullivant hold all
the patents, are entirely hand constructed, being
woven through stout galvanised iron rings in a
manner that is not to be made public property.
Mr. Pelham Bullivant is a kindly gentleman,
and it would, I am sure, grieve him sorely, if
I were to put any of you in a position to invest
your front doors with these contrivances and
get to defying your neighbours or the tax
gatherer, or, who knows, the village constable, on the strength of your immunity from
harm.

We were walking slowly back from the
wharf when I came upon a supply of wire that
had been recently landed. It looked stout and
uncompromising, and I essayed to lift a small
coil, and failed dismally.

"Surely you do not make rope with this stuff!" I remarked.

" Oh, yes," my guide replied, unconcernedly. "That is not out of the way. Now here is a piece "—singling out another coil—" of much stouter stuff—a three-eight-inch mild steel."

"Great Scott!" I ejaculated, and then I waxed facetious.

" And this, too," said I, pointing to a long, hollow steel shaft of curious construction. "Perchance you curl that up for fishing lines, or does its consistency justify you in preparing top-string from it?"

"Not quite," he replied, taking me in my humour. "We just use that out straight as a torpedo boom."

Shortly after we came into a building where the cables are brought to be fitted. This is done by hand, the cable, previously bound round with tarred rope, being doubled at the ends, and an immensely stout galvanised iron ring, termed a "thimble," inserted. It is then laid in a vice which a man screws up, while the operative pounds it into place with a heavy short-handled hammer. It is then fastened off with soft iron wire, and presents a very neat and pretty finish when the operation is concluded. Turning from this building, we crossed the tram lines again and entered the galvanising shops. Just outside I passed the acid tanks, huge structures in which caustic potash is boiled up by steam. It is in these that the wire netting to be galvanised is first of all pickled.

There was something suggestive about those tanks of bubbling, yellowish-green fluid that impelled me to keep at a safe distance. Its destructive properties are marvellous, and a thick rope hanging from a pulley above was eaten away almost past use by its action, although, as I was informed, it had only recently been put up. The men, too, who have to deal with it wear clogs, a pair of boots every three days being considered too heavy a tax on their resources as family men with other claims on their banking accounts.

Now, I don't mind confessing that I was astonished when I followed my guide into the building. There is always something that we don't know, and amongst the few matters that I could not claim to be intimate with previous to my visit is the method of galvanising with zinc.

My idea—the little thought I have ever given the subject—was that they hitched an electric

MAKING TORPEDO NETS

battery to it somewhere or other, and smeared something on it, and the job was done, being

really simple, and not particularly interesting
to watch or record.

In reality it is a sort of wholesale soldering
process. The room was as hot as a furnace,
and the heat, I soon noticed, proceeded from a
series of low, squatty coke furnaces, in the
centre of which great tanks of spelter, each
holding from eight to nineteen tons of metal,
were simmering. At the back of the one I
inspected the wire was running off a series of
bobbins through a bath of muriatic acid, which
is contained in a slate tank set just in front of
the furnace. Some of the acid splashed over
on the foreman's hand while he was explaining
the process to me; but he wiped it off quite
unconcernedly and went on talking.

"Doesn't it burn?" I queried, in astonishment, knowing something of the peculiarities
of these fluids.

"Not if you are used to it," was his
reply.

I was silenced, but not convinced, and I was
about to practically display my scepticism by
dipping my hand in the bath when Mr. Selby,
observing my action, stopped me.

"Wait a minute," he said. And he picked

up a piece of broken wire and stirred the seething liquid with it.

A dense white cloud spread out from it, and when he withdrew it I could see that it had fused. I shivered, despite the temperature, and resolved that henceforth trust should be my motto. It is well to possess an inquiring mind, I remembered; but where practical experiments are concerned it is better to temper inquiry with discrimination in large proportions. The wire passes through a heap of heated " breeze " (powdered coke), and then enters the spelter bath, being carried and kept below by a series of rollers. It emerges at the other end, and after passing through another bed of breeze to prevent oxidisation, it is coiled up in glistening white threads on another series of bobbins.

They keep the spelter at between 850 and 900 degrees Fahrenheit, and as I before remarked, the building was warm, not to say sultry. And yet there was no fire to be seen anywhere until the foreman picked out a loose brick from many such that were set at intervals around the stove, and then I could see a fiercely glowing mass that blinded me to look in upon.

Everything was warm in that building, and I gathered that the men had never raised any objections on the score of feeling chilly at their work, and were not given to petitioning for rugs or draught-excluders, or such-like vanities. Seeing the foreman test the wire by rubbing his finger along it, I did likewise. It was beautifully smooth, but it was also magnificently hot, and I spent the next few minutes in a can-can round the room with my finger in my mouth, and my guide and the foreman smiled audibly.

Wire-netting is galvanised in precisely the same way, except that a couple of men stand by it on either side with hooks in their hands as it emerges, and keep it straight and even as it is rolled up.

In a little chamber adjoining, the beautiful lesson of economy was being illustrated, the spelter being separated there in a small copper from the "hard," and formed up into cakes to be used over again. Another curious process was that of "sweating" the ashes. I have already mentioned the fact that a considerable amount of coke-dust is used in the galvanising process to heat the wire before it enters the

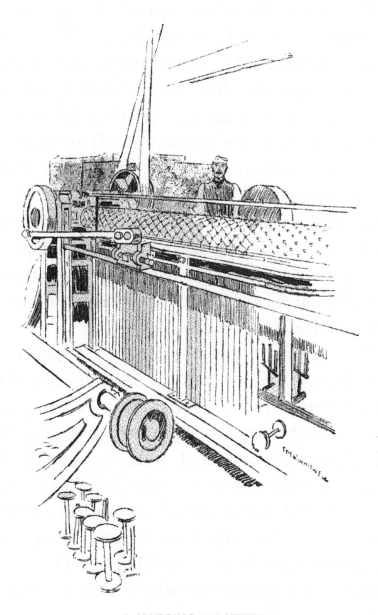

A NETTING MACHINE.

bath, and at the other end of the process to
prevent the wire from oxidising. This coke-
dust absorbes the zinc fumes, and when a
sufficient quantity has been collected, it is put
into an oven, and the zinc is extracted from it;
even then a certain amount is left in what are
called zinc ashes, which I was told are worth
about seventy shillings per ton. They do all
things on a magnificent scale in Wire-Land,
and use up spelter lavishly. A little supply of
ninety tons kept temporarily to go on with was
casually shown me stacked in plates in one of
a number of store-rooms.

The motive power—and a considerable
quantity of it is required—is derived from three
Lancashire boilers and two pair of compound
engines with widely separated cylinders, each
of a hundred indicated horse-power. I was
told that the horse-power was indicated, which
was as well, for I should never have seen it
for myself; the engine-room was too gloomy
at the moment to see anything except the dim
outline of a ponderous wheel and a maze of
intricate machinery.

I came across a fact which pleased me
excessively in connection with the engine-

house. They dodge the water company in Wire-Land, their supplies being obtained from a well in the vicinity of the boiler-house. It is pumped up and softened and purified in special tanks, and sent direct into the boilers. I have a vendetta against all water companies, hence my fiendish exultation.

After making a brief inspection of the electric lighting plant, which is contained in a small chamber opening from the engine-house and run by an engine of its own, I followed my guide out through the rope-making room to a well-stocked fitting-shop, where every appliance necessary for making and retaining in repair the immense quantity of machinery used by the firm is kept, and kept pretty continuously busy, too.

"Are you also blacksmiths and engineers?" I queried of Mr. Selby. "Oh no," he said, "we only do our own work. But we make the appliances we sell in connection with our wire rope, such as reels, nippers, thimbles, and pulley-blocks and similar work, you know." This was in addition to the making and repairing I previously referred to, and I ceased to wonder at finding six blacksmith's fires at work; and there seemed a use, too, for considerably over a hundred men and boys on the various lathes, planing, shearing, drilling, moulding, and other machines and engineers' tools, including a heavy steam hammer with which the blacksmiths seemed to revel in creating an unearthly din.

And then I came to the last scene of all, which embraced several scenes—the wire-netting departments.

Just to whet my curiosity, Mr. Selby first led me through the store-rooms, where a thousand tons of wire-netting, of assorted sizes, is kept stocked, and after I had duly pondered on the gravity of being responsible for a thousand tons of wire-netting, we went upstairs together and witnessed the process of construction.

Well, I have seen machines of every conceivable sort; I have seen many that could rival it in power, speed, and size; I have heard a few, not many, that could give it points in the matter of noise; but for sheer ingenuity, commend me to the wire-net making machine.

This machine is in operation in many shops; but I will confine myself to a description of a representative specimen of its class. It is, in the first place, a vertical machine, and like the countryman's horse, is nothing much to look at, but a marvel to go. Half the wires used in the net, each contained on a bobbin, are set underneath and lead up and into the machine by the side of a row of metal tubes, into which the other half of the wires are inserted in tightly-closed coils. A row of levers work with an oscillating motion, right and left, and the diagonal-patterned mesh is made by this motion, which carries the wire strands alternately to right and left, and fashions that twist that will be seen by examining the specimen in your own fowl-yards, presuming you keep fowls, or in your next-door neighbour's, if you don't. He will be sure to—next-door neighbours always do keep fowls where you make a special point of not doing so yourself on account of an inherited dislike of the cackling and crowing.

A wooden roller at the top, studded with spikes, engages the net as it is made, and pulls it forward to be rolled up at the back.

One gigantic machine used for constructing big-mesh nets for sheep runs of a very stout wire was capable of the most extravagant performances.

Indeed, the machines I saw could make between two hundred and two hundred and fifty miles of netting a week. I sought to discover from my guide what became of it all.

" Rabbits and pigeons? " said I, tentatively.

" They help, I daresay," he replied. " But, with the exception of a few million yards we make for the London factors, it chiefly goes to Australia. A year or two ago, we had over nine thousand miles on order for this market booked in lines of from one hundred to two thousand miles."

And so I came away back through Wire-Land, staying a while to visit the testing-shop, where every coil of wire on being received, and every rope and cable before being sent out, is thoroughly and practically tested. The appliances here vary from ingenious little contrivances, well calculated to detect faulty wire, to powerful hydraulic apparatus, which can easily put a hundred-ton strain on a cable.

And so it came to pass that I had wandered the length and breadth of Wire-Land for the space of three and a-half hours, and I had acquired for my pains much useful and interesting information, a fearful headache, and a grime-and-oil cast of countenance that I could have matched against a professional engine-driver's.

However, such things, you know, must be at every famous victory; and I thanked my guide for his courteous and most painstaking assistance, and went home in a third-class smoker, and by back ways and devious routes on getting into my immediate neighbourhood, in case anybody should see me and hail me for a chimney-sweep taking a day off, and order me to come in and sweep the copper-chimney at six the following morning.

I ONCE read a poem about a village constable who set out one morning to serve a batch of summonses on several defendants living in various parts of the suburbs of London. He wrestled with his task manfully, so the poem said, but it was a weary business, for,

. . . when he came to Gloster Road, it almost made
     him sick
To find the next defendant kept a pub. at Hackney Wick.

I know now why that constable was so overcome at his discovery. Hackney Wick is a place where it is well-nigh impossible to discover anything in the shape of an address. In fact, you speedily come to think you have good reason to congratulate yourself if you succeed in finding Hackney Wick. Nobody knows where it is if you ask, or else they have sworn never to divulge the secret, and put you off with misleading explanations and directions. They did this to me when, in the glare of the noonday sun, I left Hackney Station, and set out for Sweet-Land. I was directed to go up one street and down another, and take the third turning to the left, and keep round by a boot-shop, and when I got there some one else would tell me to go right back to my starting-place and strike out in an entirely opposite direction. I did this for nearly an hour, and then I felt like that village constable—almost sick. I hailed a cabby at length, as I stood faint and weary under the railing of Victoria Park.

"Hackney Wick, cabby," said I, "and put the pace on a bit—I'm in a hurry."

"Why, you're in it now, sir," said the cabby, looking surprised.

"In what?" queried I. "In a hurry?"

"No, sir; in 'Ackney Wick," said he.

And so it proved, and five minutes later I entered Sweet-Land by a side door and made my way up to the general offices, where I found Mr. Horn, the managing director, waiting for me.

Sweet-Land is spread out along the bank of a canal, with a large tract on the opposite side and connected to it by means of a substantial bridge, and many other ramifications in the near vicinity. Anything in the nature of a systematic exploration of it, I pretty soon discovered, was out of the question in the time that I had at my disposal. Its operations are so many and so distinctive that I walked through it for over four hours and only grasped the main processes, and even those not at all in detail. Needless to say, it is a fairly representative, and I believe it is the largest, works of its kind, being worked on the profit - sharing system and employing many hundreds of hands —the majority of them feminine; and the owners of it all stand as

CLARKE, NICKOLLS, & COOMBS, Limited,

Mr. Horn communicated these facts to me as we made our way downstairs to the showrooms, where the various kinds of confections are set out in cases for the inspection and assistance of customers.

Things were going swimmingly in Sweet-Land, I judged, and the laying down of additional boilers and plant and the erection of new buildings spoke elequently of anything but trade depression.

An immense quantity of steam is required for boiling purposes as well as for motive power, and the boiler accommodation has to be continually increased to meet the growing demands.

"We are rather slack here just now," said my guide, leading me into a long, open room of tropical temperature.

We were in the jam-making department I saw then, and I understood that I had come at an inopportune time. The fruit seasons influence this department, of course, and although a certain quantity of jam, of one description or another, is always being made, and marmalades serve to keep the building in a state more or less active all the year round, I could then only witness the nature and not the extent of the processes.

The raw materials, mostly sugar and fruit I should premise, come into Sweet-Land by way of the canal previously alluded to, and in vans direct from the fruit-growing counties.

The jam-making departments consist of a series of large rooms on the ground floor, set about with rows of shining copper boilers, in which the fruit is boiled up with sugar by steam, being stirred continually the while by means of a rod with a blade on either side, actuated by a cog set over each pan. From these pans the jam is tipped out into copper store tubs, each provided with a very accommodating tap, through which it flows into smaller cans that serve to make the bottling or " jarring " process simple and easy.

Away in a further corner I came upon a contrivance for dealing with stone fruits, in the shape of a big, horizontal, copper-lined box, in which the plums and cherries and anything else requiring treatment are rubbed through a copper sieve into a chest below, the stones being retained and cleared out by the sackful.

Opposite this a hydraulic press of immense power for pressing fruits was stationed; but

BOILING THE SUGAR.

this, like the other machine, was just then idle,
although in a month's time its services will
again be in great demand.

Through the further door I came upon a still
used in the preparation of essences for flavouring purposes, and in the essence macerating
room adjoining, lemon and other essences were
stored in a series of sixty-gallon drums. A
businesslike-looking laboratory for testing purposes completed this branch of Sweet-Land,
which has a temperature so high that I found
it convenient to go and sit in the boiler-house
awhile to cool off a bit before continuing my
exploration. The alleviation I thus obtained
was immediately afterwards discounted, for,
closely following my guide, I found myself in
a building in which numerous coke fires were
boiling up huge coppers of sugar at a temperature of 320 degrees. I began to ooze and
decline visibly as I stood between them while
Mr. Horn initiated me into the mysteries of
making the succulent "drop." The process is
simple but ingenious. The sugar is first melted
with water, and then boiled up to the required
temperature, when it is poured on to iron
tables, and kneaded out into long flat slabs.

While these are still warm and pliant they are passed through heavy gun-metal rollers, which are stamped out to form a series of half-moulds of the shape the drop is to assume. Two halves of these moulds meet together as the rollers revolve, and when the slab has run its length through them, it only has to be set to get cold, when a knock will break it up into more drops than half-a-crown would purchase retail at the " sucker shop."

Affixed to each of the numerous pillars that run from floor to roof up and down the room is a stout iron hook, and these, I was shown, are used for sugar drawing. A workman will roll a mass of the pliant dough into a round length, and will cast it over the hook and draw it out with an astonishing rapidity. The effect of this manipulation is soon observable in the lightening of colour, and after a time it becomes quite white and opaque, being at the outset of a pretty red tinge.

As any schoolboy is aware, these drops are of all colours and shapes and sizes, and are flavoured according to the name they bear.

I was very much struck with the cheerful docility of these sugary compounds under treat-

ment; and the good-natured way the drops
consented to separate themselves wholesale in
response to tapping, instead of insisting on
being pulled apart by tedious hand labour,
affords a lesson in docility and wisdom to any
citizen who sets his back up and struggles
vainly but obstinately with the authorities for
his rights.

Cocoanuts play a very large part in sweet-
stuffs, and one building I found given over
solely to preparing them for future treatment.
Huge sacks of nuts were stored up there waiting their turn to go under the small-toothed
circular saws which sever the shells into four
equal quarters and leave the nut entire.
This is managed by a gauge that prevents
the blade biting too deeply, and the rapidity
with which the operation is carried on by
girls is astonishing, or it would be to those
who have not come to understand how very
perfect long continued practice can make one.
If I had to do a job of that sort I should take
it much more leisurely, at the outset at any
rate, for I should attach a greater importance
to keeping my fingers whole and entire.

Another ticklish operation from my point of

MOULDING SWEETS.

view consists in removing the husks, which is also done in machines where a series of rasps revolve at the rate of 2,000 revolutions a minute. "Gr-gr-gr" it went, and the young lady whose privilege and pleasure it was to exhibit her skill before me handed me a nut from which the brown husk had been evenly and completely removed.

After that the nuts are thrown wholesale into the feeding trough of a grating machine, and fall below it in a white powder suggestive of a snowstorm in miniature.

Wandering through rooms filled with feminine energy and dexterity, I came upon another endless array of steam-boiled coppers, where incipient sweetstuff hissed and bubbled as it was stirred vigorously round by machinery. A double motion is ingeniously imparted to these revolving blades by having one set loosely and actuated in a contrary direction by means of a cog-wheel.

Ascending from the heat and the clatter below, I entered a room calculated to make an ordinary man of small means feel sick. It is where certain of the flavouring essences are kept, and on a series of unimposing-looking shelves, in jars and bottles and small metal tanks, the value of thousands of pounds is stored up in the shape of vanilla and other costly extracts, oils, and essences.

"Concentrated banking," you might term this means of laying up treasure; although if I had a few hundred pounds lying idle I think I should still prefer the old-fashioned method of storing it up against a rainy day. Banks do break occasionally, but not so readily as glass bottles do; and to have £70 set up on a shelf about the house in a twelve-ounce phial would, I feel sure, be a source of anxiety to me that would prove a serious obstacle to the maintenance of a calm and placid frame of mind.

Of course, these essences are not used just as I saw them. They certainly form a very considerable item in the expenses of Sweet-Land; but that would be altogether too costly. As a matter of fact, they are first diluted down in spirits, and several cans in process were even then ranged round the room in which the essences are stored.

A great quantity of the essence is used in the preparation of table jellies, of which many tons are turned out from Sweet-Land annually.

Continuing my travels, I entered still another wrapping and packing room, and in the corner of it came across a waxing-machine for preparing the wrappers. Long sheets of paper are run through a bath of melted wax, and are passed through a pair of powerful rollers which squeeze the superfluous fluid from them, and leave them dry and ready for use. The thinner paper which will be found round caramels and similar sweets is bought, it being found cheaper to obtain it thus than to prepare it on the premises.

I think, without exception, Sweet-Land is the most intricate of any that I have explored in the way of business. Devious turnings, unexpected staircases, and rambling passages led me through more rooms than I could take count of, in which hundreds of busy girls and women were employed in various preparing and packing operations, in which prize-packets, bonbons, and sherbet were bottled and boxed with marvellous dexterity.

Through a long, narrow passage, and up a supplementary staircase, which usually serves as a provision in case of fire, and constrained me to duck my head and go warily in the

STEAMING SWEETS.

interests of my head-gear, and I emerged on
something definite—a distinctive process that
I could lay hold of and follow.

There was warmth, and a pleasant smell
pervaded the atmosphere with a subtle infusion
of something in it that promptly resolved itself
when, in answer to my query, Mr. Horn
remarked tersely : " Ginger conserves."

The ginger is imported in immense casks in
syrup, and is shot out, as the first step towards
its preparation, on to sieves to drain. It is
then cut up into squares with dexterity and a
knife, boiled, and left to crystallise in huge
tanks.

Taking a casual view of the export packing
department in an adjoining chamber as I
passed through it, I came upon a very
interesting process—a sort of sweet-foundry,
where fishes, and frogs, and other saccharine
novelties are moulded.

Everything within that room, including the
operatives, was white as white could be, and I
began to take on the prevailing hue before I
had been there two minutes. The moulds are
formed in starch, long narrow tables covered
with this material standing down the room in

serried rows. The matrix is formed by a board, on which dozens of the articles to be moulded are carved in relief, it being the duty of one set of operatives to walk down the table and impress this firmly in the receptive starch-bed. Other girls follow up behind with curiously constructed cans of liquid sweetstuff, having anything from six to a dozen spouts set at proportionate distances, and by a dexterous movement tip a supply of its contents into the moulds.

The starch is subsequently brushed or blown off the sweets by hand, and in the case of gum drops and such-like varieties, they are further cleaned by a steam jet.

It is a necessity and a great convenience in Sweet-Land, owing to its numerous and varied operations, that each department shall be practically complete in itself, and worked through, from boiling and preparing the material to packing the finished articles, in its precincts. There are certain exceptions to this rule, of course, but as a general thing it is adhered to.

When I grasped this fact I came to look for the rows of shining copper pans as a matter

of course—I should have been disappointed if
I hadn't seen them; and the moulding rooms
are no exception to the rule. There they stood
ranged round the wall, and differing in little
except the temperature at which they are boiled.
This difference is not perceptible to the naked
eye, and is only apparent in the rows of steam
pipes that the peculiarity necessitates. If
sugar were not so finnicky in this particular,
one main steam supply would serve the purpose
that now necessitates almost a separate pipe to
each pan; but this is one of the rights that
sugar is inclined to stand up and fight for, and
they deem it best to humour it and to supply
steam at high or low pressure according to
requirements.

A light iron staircase runs from ground to
roof on all the buildings in Sweet-Land, with a
landing on each floor. The main purpose of
these is to meet the yearning desire to reach
the ground that would doubtless arise in the
breasts of the operatives should a fire inoppor-
tunely break out below. I know I should feel
particularly grateful for the kindly forethought
if I found myself in such a plight. As it was,
I appreciated it when, after a prolonged tour

IN THE LOZENGE ROOM.—STAMPING LOZENGES.

through extra-heated rooms that hot May
day, I debouched on to it in company with
Mr. Horn, and took a breathing spell in the
pleasant breeze that blew across the canal
below me.

Not for long, however, could I view the rural
charms of Hackney Wick that were spread out
so alluringly around me. Duty called, and to
that voice let no man, woman, or little children,
so that they value their well-being, turn a deaf
ear. I never do; but if I chance to be too
busy, or too lazy, or too tired, and deem the
call untimely, I say: "Oh, yes! I hear you
quite plainly, but I ain't a-coming."

In this case 1 did go, and after following the
roof round I entered another building, where
liquorice is prepared for the market.

Huge tubs filled with the black paste were
standing round, and from these the contents
are supplied to the cooking-kettles.

The liquorice is made up in various styles,
from big hollow tubes to the long black strips
that pass with schoolboys as " shoe-laces."

Some ingenious and complicated machinery
is concerned in preparing the liquorice for the
trade, and one of these expresses a row of six

continuous long thin black tubes on to a board that is put to receive them by the woman in charge. They lie side by side like policemen detailing an assault case, and adhere to each other, though what they were intended to represent I couldn't conceive. Doubtless it was merely a schoolboy's whim in preferring his liquorice fashioned to receiving it in square chunks.

A neat and noisy little machine is also at work on this floor husking almonds, which is done wholesale, the nuts being fed into the top and thrown out below with every shred of skin removed.

And then we went downstairs, and upstairs, and round corners, and through passages, and came at length into the lozenge-making department, where some novel features were to be encountered.

Great masses of dough, containing the required ingredients—soda, sulphur, or cough-cure, as the case may be—mixed in with the sugar or gum solution, are first prepared and rolled out into long flat slabs. These are thrown on to a travelling platform which carries them under a series of punches, all working together, stamping out lozenges by the hundred. A feature in this machine is that the network of lozenge material left after the punching process is carried back underneath to be made up again and once more passed through the punches. It is entirely automatic in action, and deposits the lozenges on a travelling blade, which carries them forward at each movement and throws them out on a table to be gathered up and added to the general stock.

We went outside again after leaving the lozenge department, and, descending to a lower floor, my guide opened a door, and my ears were saluted with an awful din. Two cyclones, one thunderstorm, and a small boy with a tin can and a tack-hammer might rival it if they worked hard; but it would be a task.

The sight I obtained when I got inside, however, was the most striking and impressive of any that I have witnessed in connection with these explorations of mine. A number of immense copper pans termed "comfit-pans," capable of holding many hundredweight of material, were set diagonally and, revolving

THE COMFIT PANS.

slowly, kept in motion bushels upon bushels of
sweets. There were sixteen of these pans in

all, and I gathered from Mr. Horn that they are used for all spheroid-shaped sweets. The pans contain a false convex bottom, which is used as a steam chest, and the sweets are built up from seeds—like aniseed, for instance, in the case of the familiar aniseed balls—and almonds, by a continuous supply of boiling syrup, which is thrown upon them from time to time by the men in charge. The motion distributes this fairly and evenly around them, and the seed gradually increases in size while retaining its original shape, until it presents a serious obstacle even for the average schoolboy to tackle.

The process is slow, but very sure, and as an instance of what the comfit-pans are capable of effecting Mr. Horn informed me that four hundredweight of some minute seed, such as caraway or aniseed, rolled round in them for two months, will result in twenty tons of sweet-stuff. The pans are emptied every night, and the contents are kept in a series of heated chambers ready for the morning's resumption of work.

The next time a moralist seeks to floor me with that silly little fact about a rolling stone not being good at gathering moss—and it may be because there is no moss to gather—I shall go him one better, and inform him that a rolling caraway-seed at any rate can gather a considerable quantity of sugar.

"This is our main supply of material," remarked my guide a few minutes later, as we turned into another building that was white from floor to roof with the concentrated essence of sweetness.

It was, I discovered, the sugar-grinding room, where an insignificant-looking machine of marvellous possibilities disputed the floor space with an immense chamber constructed of wooden staves and a very close netting.

It takes a special engine, and one of considerable power too, to operate the crusher in which a drum, set with short, stout spokes, revolves at 3,700 revolutions a minute, and quickly converts the sugar into a dust as fine or finer than cornflour.

It is difficult to form an idea of what this really means, and I myself only understood it by outside considerations. For instance, a capacious copper pipe leads from the crusher to the receiving chamber, and in obedience to

Mr. Horn's suggestion I laid my hand on it. It was hot, and, I was informed, was made so solely by the friction of the sugar which is blown through it by the velocity of the revolving drum of the crusher.

"That is where we make caramels," said Mr. Horn, indicating a jealously closed door that opened out on our right; "but——"

"A trade secret?" said I.

Mr. Horn nodded, so I passed caramels and went to see chocolate instead.

In a series of immense rooms, filled with gleaming machinery of complicated and ingenious device, I found this operation being carried on in sections. After the cocoa bean has been roasted by gas, it is put into a coffee-mill sort of affair to be crushed, the cocoa being separated from the husks and falling to the floor below. After that it is put through a mill which has a steam-heated chest in its construction, and then as a result of the heat and pressure the cocoa flows out in a thickish oily liquid, which is caught in cans set purposely below.

I daresay you have heard that old chestnut about King Georgey and the apple-dumpling!

PACKING CHOCOLATES.

Well, I was reminded of it when I saw them making chocolate creams, which, by the way, is merely sugar whipped up into a creamy white paste by a revolving blade, in pans very similar to those used in sweet-making. I have often mentally admired the patience and perseverance expended by the sweet-makers in filling yearly millions of these little chocolate cases with the cream, and now I have learnt, as a matter of fact, that they don't do anything of the sort. They put the chocolate over the cream wholesale by a dipping process, the cream being first fashioned and then thrown into pans of liquid chocolate. After this it merely remains to fish them out again, and lay them to drain on wire gratings.

It was about this time that I discovered, to my dismay, how late it was growing, and the remainder of my exploring was rather in the nature of a hasty scramble. I lingered longest in the butter-scotch rooms, for the operation recalled familiar scenes of my youth, when I would borrow cook's saucepans, without leave, and fill them with a messy concoction of butter, sugar, and bits of orange or lemon peel, and then burn the bottoms out in endeavouring to cook it, unless Martha happened to appear in time to box my ears and discourage me from my enterprise.

Art in sweet-making was illustrated in several pretty and effective specimens of mosaic work, designed and carried out by a deaf and dumb operative, quite a young fellow, too, who makes this a speciality, and of whom the firm are rather proud.

I mentioned that Sweet-Land spreads out on both sides of the canal, and, on crossing the bridge, I found that candying peels and box-making are the staple industries in the branch over the water. I came to the peel-yard first, and discovered huge tubs of halved oranges and lemons, which are kept in salt water to preserve them until they are required for treatment. When this time draws near, they are steeped in fresh water that is continually being changed, and pulped.

It is necessary to retain the pulp in them to the last minute, I gathered from Mr. Horn, to preserve the skins in shape, and when the interior is removed they are transferred to tubs of syrup, from which they emerge, transparent and tempting, to aid in the manufacture of cakes and cookies.

It is rather a reversal of the general order of things to retain the skins and throw the pulp away, and one of these days I shall found a moral story on it about the uppish orange and the triumphant peel.

The stables are also situated on this side of the canal, and are designed and fitted with commendable care and thought for the reception of the firm's dumb *employés*.

Immense wood-piles, too, encumber the ground, and after making an inspection of the extensive and well-appointed box-making buildings—where tools and machines are alike, with few exceptions, worked by girls with a skill that discounts the musty old jokelet about women not being able to wield a hammer—I gained an increased respect for the extent of the firm's business, more particularly when my guide informed me that in addition to their own manufacture quite one-third of the sweet-boxes used, and all the packing cases, have to be imported ready-made. Cardboard boxes are made in another department across the road, where German confectionery is prepared, in separate buildings that were once blocks of private houses, and are now a provoking and moral-destroying maze to any one who endeavours to find his way round in them unattended.

I visited these departments later, and saw Easter eggs and Christmas-tree decorations, tasteful (using the word in its double sense) little sweet baskets and bird-cages, being prepared, it being the constant endeavour there to devise fresh and attractive novelties for each· recurring season.

Easy as these processes are to catalogue, they involved the expenditure of much energy and time to inspect and record, and I have an uneasy feeling that I am not doing sufficient justice to my subject, even when I add that table jellies, essences and fruit cordials for summer and winter drinks, and sauces of various descriptions form part of the operations in Sweet-Land, and were all duly shown and explained to me freely and with no unnecessary reservation.

But I am generally considerate to these gentlemen, and I am always so towards myself, and it was fearfully late and I was travel-stained and weary, so that when my guide hinted at further explorings I struck, in his interests and my own.

Mr. Horn's information—and no one is better able to give it, seeing that he has watched the place grow round him for the past twenty-three years—seconded my impressions, obtained by practical inspection, that Sweet-Land and its proprietors merit every commendation—firstly, as a firm; secondly, for its manufactures; and lastly, but not least, for the instructive and entertaining qualities it presents to a perambulating scribe such as myself, with a big streak of professional and personal curiosity in his composition. I give the commendation on all counts as heartily as I appreciate the courtesy and attention which I was accorded on my visit.

A small boy undertook to guide me back to the railway-station for a coppery consideration, and he opened his mouth wide in wonderment when I talked to him, as the Ancient Mariner did to *his* victim, of tons upon tons of sweet-stuff, and babbled musingly of square miles of candied peel and yards of liquorice. He said I was a " corf-drop " when I gave him his twopence. I know I was sugar-coated, and a sticky sweetness clung about me for days afterwards.

LONDON:
W. SPEAIGHT AND SONS, PRINTERS,
FETTER LANE.

UNIVERSITY OF CALIFORNIA LIBRARY
Los Angeles
This book is DUE on the last date stamped below.

Form L9–50m-4,'61(B8994s4)444

Lightning Source UK Ltd.
Milton Keynes UK
UKHW012051230219
337878UK00014B/1288/P